US Military Innovation since the Cold War

This book explains how the US military reacted to the "Revolution in Military Affairs" (RMA), and failed to innovate in its organization or doctrine to match the technological breakthroughs it brought about.

Many called for the transformation of the US military in the years after the end of the Cold War, seeking the changes in organization and doctrine that would complete the RMA innovation and a commitment to counter-insurgency, peacekeeping and nation building missions. This volume describes the origins, uses, and limits of the RMA technologies, examines how each of the five US armed services (categorizing the Special Operations as a separate service) made their adjustments both to the technologies and the use of force, and how the role of the civilian officials and the defense industry altered in this process of change and avoidance of change.

The book examines the internal politics of the services as well as civil/military relations to identify the external pressures on the services for significant change in their doctrine and weapons. Many have noted the failure of the services to innovate in what can be called the "Second Interwar Period" (the years after the Cold War). This book offers explanations for this failure and arguments about the possible range and desirability of military innovation in the post-Cold War era.

This book will be of great interest to students of strategic studies, US defense politics, military studies, and US politics.

Harvey M. Sapolsky is Professor of Public Policy and Organization in the Department of Political Science at MIT and former Director of the Security Studies Program. **Benjamin H. Friedman** is a Research Fellow in Defense and Homeland Security Studies at the Cato Institute and a PhD Candidate in Political Science at MIT. **Brendan Rittenhouse Green** is a PhD Candidate in Political Science at MIT and an affiliate of the Security Studies Program.

Series: Strategy and history
Series editors: Colin Gray and Williamson Murray

This new series will focus on the theory and practice of strategy. Following Clausewitz, strategy has been understood to mean the use made of force, and the threat of the use of force, for the ends of policy. This series is as interested in ideas as in historical cases of grand strategy and military strategy in action. All historical periods, near and past, and even future, are of interest. In addition to original monographs, the series will from time to time publish edited reprints of neglected classics as well as collections of essays.

1 **Military Logistics and Strategic Performance**
Thomas M. Kane

2 **Strategy for Chaos**
Revolutions in military affairs and the evidence of history
Colin Gray

3 **The Myth of Inevitable US Defeat in Vietnam**
C. Dale Walton

4 **Astropolitik**
Classical geopolitics in the space age
Everett C. Dolman

5 **Anglo-American Strategic Relations and the Far East, 1933–1939**
Imperial crossroads
Greg Kennedy

6 **Pure Strategy**
Power and principle in the space and information age
Everett C. Dolman

7 **The Red Army, 1918–1941**
From vanguard of world revolution to US ally
Earl F. Ziemke

8 Britain and Ballistic Missile Defence, 1942–2002
Jeremy Stocker

9 The Nature of War in the Information Age
Clausewitzian future
David J. Lonsdale

10 Strategy as Social Science
Thomas Schelling and the nuclear age
Robert Ayson

11 Warfighting and Disruptive Technologies
Disguising innovation
Terry Pierce

12 The Fog of Peace and War Planning
Military and strategic planning under uncertainty
Edited by Talbot C. Imlay and Monica Duffy Toft

13 US Army Intervention Policy and Army Innovation
From Vietnam to Iraq
Richard Lock-Pullan

14 German Disarmament After World War I
The diplomacy of international arms inspection 1920–1931
Richard J. Shuster

15 Strategy and History
Essays on theory and practice
Colin S. Gray

16 The German 1918 Offensives
A case study in the operational level of war
David T. Zabecki

17 Special Operations and Strategy
From World War II to the war on terrorism
James D. Kiras

18 Science, Strategy and War
The strategic theory of John Boyd
Frans P.B. Osinga

19 US Defense Strategy from Vietnam to Operation Iraqi Freedom
Military innovation and the new American way of war, 1973–2003
Robert R. Tomes

20 US Special Forces and Counterinsurgency in Vietnam
Military innovation and institutional failure, 1961–63
Christopher K. Ives

21 War in Iraq
Planning and execution
Thomas G. Mahnken and Thomas A. Keaney

22 Alexander the Great
Lessons in strategy
David J. Lonsdale

23 America, Technology and Strategic Culture
A Clausewitzian assessment
Brice F. Harris

24 US Military Innovation since the Cold War
Creation without destruction
Edited by Harvey M. Sapolsky, Benjamin H. Friedman and Brendan Rittenhouse Green

US Military Innovation since the Cold War

Creation without destruction

Edited by Harvey M. Sapolsky, Benjamin H. Friedman and Brendan Rittenhouse Green

Routledge
Taylor & Francis Group
LONDON AND NEW YORK

First published 2009
by Routledge
2 Park Square, Milton Park, Abingdon, Oxon OX14 4RN

Simultaneously published in the USA and Canada
by Routledge
605 Third Avenue, New York, NY 10017

Routledge is an imprint of the Taylor & Francis Group, an informa business

First issued in paperback 2012

Typeset in Times by Wearset Ltd, Boldon, Tyne and Wear

British Library Cataloguing in Publication Data
A catalogue record for this book is available from the British Library

Library of Congress Cataloging in Publication Data
A catalog record for this book has been requested

ISBN 13: 978-0-415-62230-1 (pbk)
ISBN 13: 978-0-415-77791-9 (hbk)

To Paris Genalis, senior defense civil servant and proud naval architect who was our inspiration and supporter

Contents

Illustrations

Figures

Tables

Contributors

David T. Burbach is a Professor at the US Naval War College. Prior to that, he was a Professor at the Army's School of Advanced Military Studies at Fort Leavenworth, Kansas. His research focuses on US public opinion on war and diplomacy and nuclear weapons. He holds a PhD in Political Science from the Massachusetts Institute of Technology.

Benjamin H. Friedman is a Research Fellow in Defense and Homeland Security Studies at the Cato Institute and a PhD Candidate in Political Science at MIT and an affiliate of the Security Studies Program. His research deals with American defense politics and threat perception.

Eugene Gholz is an Associate Professor at the LBJ School of Public Affairs at the University of Texas in Austin. He studies innovation, defense management and organization, and US foreign military policy. He is the co-author of *Buying Military Transformation: Technological Innovation and the Defense Industry* (with Peter Dombrowski) and of *US Defense Politics: The Origin of Security Policies* (with Harvey M. Sapolsky and Caitlin Talmadge). His recent work has emphasized the topics of systems integration and energy security. He is also a Senior Fellow of the Robert Strauss Center on International Security and Law, a Research Affiliate of the MIT Security Studies Program, and Associate Editor of the journal *Security Studies.*

Brendan Rittenhouse Green is a PhD Candidate in Political Science at MIT and an affiliate of the Security Studies Program. His research focuses on American liberalism and US grand strategy.

Colin Jackson is a Professor at the US Naval War College, holds a PhD in Political Science from MIT, and is an affiliate of the Security Studies Program. His current research deals with counter-insurgency, organizational learning, and intelligence operations. He served four years with the US Army as a tank and cavalry officer and continues to serve as a military intelligence officer in the Army Reserve. He has an MA from the School of Advanced International Studies at Johns Hopkins University and an MBA from the University of Pennsylvania.

Austin Long is an Adjunct Scholar at the RAND Corporation, a PhD candidate in the Department of Political Science at MIT and an affiliate of the Security Studies Program. His research deals counter-insurgency doctrine, deterrence, intelligence operations, and organizational politics. He holds an MA from the Georgia Tech in International Affairs.

Harvey M. Sapolsky is Professor of Public Policy and Organization in the Department of Political Science at MIT and former Director of the Security Studies Program. His current research focuses on three main areas: US defense politics including especially inter-service and civil/military relations; the impact of casualties on the use of force; weapon acquisition policies, military innovation, and the structure and performance of defense industries. Professor Sapolsky holds a MPA and PhD from Harvard University. He is the author or editor of eight books on the politics of defense and health care.

Robert R. Tomes is President of Liminal Leadership, LLC and is a Lecturer in Security Policy Studies at the George Washington University. He is the author of *US Defense Strategy from Vietnam to Operation Iraqi Freedom* (Routledge, 2007), the co-editor of *Africa at the Crossroads* (CENSA, 2007) and the author of numerous book chapters and articles. He holds a PhD in Political Science from the University of Maryland.

Sanford L. Weiner is a Research Associate at the Security Studies Program, at the MIT. He studies organizational change and innovation in both military and public health agencies and directs SSP's summer Professional Courses on "Promoting Innovation: The Dynamics of Technology and Organizations." He has also worked at the Health Policy Center at Brandeis, the Graduate School of Public Policy at the University of California, Berkeley, and at Queen Mary's College, University of London.

Preface

This book began as an attempt to explain how structural change affects innovation. Specifically, the work was directed at specifying how the centralization of defense acquisition authority and the legislative mandate that the armed services be more coordinated (or, as it is called, "joint"), which occurred in the 1980s, affected the Revolution in Military Affairs (RMA), the changes in weapons technologies that allow targets to be identified, targeted, and struck precisely that seemed to revolutionize warfare in recent years, and most especially in the Gulf War.

Preliminary work showed the RMA was less like a revolution and more like an extension of service plans and preferences. Missing was the organizational change required for significant innovation. Several extensive discussions with faculty, graduate students, and friends of the MIT Security Studies Program gave the project a different orientation. Instead of explaining innovation, we were debating the causes for its absence. The structural changes were just one in a series of explanations that we explored.

The new focus was on the differences between technological and doctrinal change. The Second Interwar Period saw much more of the former and surprisingly little of the latter. The 1990s saw the acceleration of some changes that had long been in process and the blockage of others that seemed so desirable from a policy perspective. Understanding the politics that directed these outcomes is what the book is about. It is a product of a team effort, one that we enjoyed both for its intellectual challenge and good fellowship.

Acknowledgments

We have several individuals and organizations to thank. The Center for Technology and National Security at the National Defense University (NDU) in Washington, DC, gave sponsorship to our quest to understand recent military innovation. The Security Studies Program at the Massachusetts Institute of Technology in Cambridge, Massachusetts, gave a home to the project. It was an administrative team effort.

Two NDU staffers, Professor Mark Montroll of the Industrial College of the Armed Forces, a component of NDU, and the late Dr. Paris Genalis, initially on leave at NDU from the Office of the Under Secretary of Defense for Acquisition, Technology, and Logistics, and then a defense consultant, were active participants in the project and played vital roles in helping us understand both the relevant technologies and the relevant Defense Department processes. They were crucial in shaping the project's boundaries and insights.

Several at the MIT Security Studies Program were very helpful to us. Owen Cote gave more than he knew to our work. He is one of the nation's most knowledgeable analysts of military doctrine and the very rare security studies specialist who can handle equally well developments in military technology, strategic thinking, and organizational politics. Barry Posen sets the intellectual pace at the Security Studies Program and clearly influenced our thinking in the work reported here. We happily recognize that Caitlin Talmadge was vital to the project at all of its stages and was its fourth co-editor in fact if not in font. Robert Murray and Magdalena Rieb helped guide us through the administrative maze that is the academic research contract process. And, as always, Lynne Levine was there to help prevent us from stumbling badly as we tried to pull this complex project together and shove it out the door.

1 The missing transformation

Harvey M. Sapolsky, Brendan Rittenhouse Green, and Benjamin H. Friedman

Until recently one of the safest places to be during an American bombing attack was standing at the center of the target. Hundreds of bombers, at great risk to their crews, flew in each bombing mission during World War II in the hope that just a few of the thousands of bombs they dropped would hit the intended target. Despite efforts at calculating wind effects and release timing, most bombs fell in rather random patterns, either blowing up harmlessly or killing and wounding the innocent. In Vietnam the experience was hardly much better. This problem led to many solutions: justifications for attacking enemy civilians, the search for bigger and bigger bombs, and the quest for accuracy. The results essentially in sequence were:

1 the fire bombing of cities – doing what could be done;
2 the atomic bomb – making missing not matter by making a bigger explosion;
3 killing lots of targets accurately.

This third change is of great importance for warfare today, and calls out for analysis. Some history reveals its significance.

Concern for reducing casualties on both sides of the bomb bay doors was a crucial motivator in the push toward accuracy. The Army Air Force suffered more than twice as many aircraft crew-member deaths in the bombing campaign in Europe than did the Marine Corps in its heroic assaults on Japanese-held islands in the Pacific.[1] German antiaircraft guns, Luftwaffe interceptors, mechanical failures, and bad weather took a huge toll on aircraft and airmen in the near futile effort to slow down production supporting the Nazi war machine. The use of effective Soviet-designed antiaircraft missiles by the North Vietnamese forced the substitution of more maneuverable fighter bombers for the more vulnerable and more expensive bombers, but to limited avail because the American public's tolerance for air crew losses had declined considerably since World War II. Similarly, although civilian deaths in World War II required various military justifications, they were generally accepted by the American public, but by the end of the Vietnam War they were not. Planes and pilots were lost in Vietnam to the political requirement to avoid civilian casualties, accelerating greatly the interest in acquiring precision weaponry.

After Vietnam, the American military refocused on the challenge of stopping a potential Soviet ground attack in Europe. New equipment and doctrine were developed that incorporated improvements in sensors, accuracy, and communications. The planning was geared especially against the multi-echelon array of forces standard in Soviet doctrine and emphasized a coordinated air–land battle that featured rapid movement and maneuver and the full exploitation of American technological advantages. Soviet strategic planners followed discussions of these capabilities closely, which they dubbed the Reconnaissance Strike Complex, and despaired at the ability of the Soviet economy to match or counter.[2] The Reagan-era build-up, which rapidly recapitalized and reinvigorated the American military, seemed to confirm America's capacity to exploit fundamental advantages in terms of its economic power and technological development.[3] Confused and exhausted by its leaders' failed attempts to match both the build-up and the parallel expansion of Western consumerism, the Soviet Union fell apart.

The Gulf War revealed to the world the effectiveness of America's post-Vietnam military reformation. The liberation of Kuwait was a rather effortless campaign for the American-led coalition that formed to confront Saddam Hussein, demonstrating both the efficacy of American airpower and the overwhelming superiority of its ground forces. The role of stealth aircraft, thermal sensors, Global Positioning System (GPS) navigation, laser-guided bombs, satellite communications, night vision equipment, and radar surveillance of the battlefield via JSTARS, the ground monitoring and targeting aircraft, were all heralded in the many glowing studies of the war. As the Soviets knew, the American military had outpaced all others in terms of applying technology to conventional warfare problems – yielding a change in warfare not unlike the one that occurred before World War II.

The second inter-war period is our term of art for the time from the fall of the Berlin Wall and the attacks of September 11, 2001. The first inter-war period as recognized by military analysts took place between the two world wars, from 1919–1938. It was a time of revolutionary military innovation, witnessing the development of the tank and the doctrine of blitzkrieg; aircraft carrier aviation and the revolution in naval warfare; and strategic bombing and integrated air defense. The second inter-war period was similarly situated between a great conflict and the next, and also saw great technological upheaval. The term captures both the warfare changes many observers saw on the horizon, and the demand for innovation stimulated by a new and uncertain international environment. For many observers, this kind of innovation was redolent in the US military: America was leading a Revolution in Military Affairs (RMA) that was changing the pace and structure of warfare.[4]

Establishing the base

But did US military innovate during the second inter-war period? A review of post-Cold War history suggests that the story of the RMA in practice was not so simple. American defense policy in the second inter-war period was driven by a new and unfamiliar environment: the need to preserve budgets in the face of domestic pressures and an international system prone to chaos in areas of tertiary interest. There were technological advances, but was there innovation?

Even before the Berlin Wall came down planning began within the US military for life after the Cold War. This work accelerated when General Colin Powell became Chairman of the Joint Chiefs of Staff. He believed that the bipolar world of the Cold War was quickly passing, and that its end would lead to demands for a substantial "peace dividend." The planning he approved saw the provision of regional stability as the likely replacement for the Soviet-focused global containment strategy that the US had been pursuing. But most of the post-Cold War work at Joint Staff was absorbed with the need to head off large force structure and budgetary cuts once the realization set in that the long struggle with Communism and the Soviet Union was over.

In 1990, leaders of the Democratic Congress, led by House Armed Committee Chairman Les Aspin and Senate Armed Services Committee Chair Sam Nunn, indeed sought significant reduction in military spending, the peace dividend Powell feared. The Bush Administration, led by Powell, agreed to some cuts, especially once events made the extent of the Soviet Union's demise clear.[5] Powell believed that if the military failed to organize and manage the drawdown, it would occur anyway, without their control, leading to what some call a hollow force, where forces and missions are maintained without sufficient funds to train the forces and execute the missions. Thus, Powell argued, the military had to get out in front of Congress, to manage the drawdown.[6] The force structure changes proposed, referred to as the Base Force, were meant to be a floor below which the military was not willing to go. Each service's slice of budget was cut by a roughly similar amount, reflecting a conclusion that it would be too much trouble politically to cut the forces in any but an equal way. Even these cuts found complaints among the service chiefs who were largely bypassed in the formulation and presentation of the drawdown plan. The Navy was to go to twelve aircraft carriers and 400 ships from the existing fifteen and 500 plus, respectively. The Air Force was to drop to fifteen active and eleven reserve tactical wings, the new measure of airpower strength. The Army cut would include a loss of six to eight active divisions from the then current eighteen, and to a base of 550,000 soldiers from the 750,000 plus active force. The Marines were slated to decline from nearly 200,000 to 150,000 personnel but managed to avoid cuts below the 170,000 mark.[7]

Before any of these changes could take place, the United States, leading an international coalition, went to war to oust Iraqi forces from Kuwait. Over 500,000 servicemen and women were deployed to the Persian Gulf region for what turned out to be a quick and relatively pain-free victory. Many of the units were disbanded when they returned home. Garrisons in Europe and Asia were reduced. With only modest adjustments – taking more from the Army and less from the Marines for example – the Base Force became policy. The decade after the end of the Cold War would be one of retrenchment, but not total demobilization. US force levels were about one-third smaller in 2000 than they were in

1990, but in 2000 such forces made the US a military colossus among nations, accounting for almost half of the world's defense spending.

The many strategy documents, commission reports, and high level studies produced following the end of the Cold War left largely untouched the force levels that Colin Powell put forth as the fatal cracks in the Berlin Wall appeared. Meanwhile, a series of international crises developed in the wake of the collapse of the Soviet Empire, beckoning various American responses, from peacekeeping and humanitarian relief operations in consort with allies to unilateral combat strikes. The military's force structure, though, remained largely unconnected to these missions. Through it all America's grand strategy remained muddled, as did the rationales for particular numbers. Why we needed 300, 313, or 425 warships was very difficult to explain. The same was true for particular numbers of divisions or aircraft wings. Nevertheless, the Base Force selections persisted with attempts at even minor adjustments causing alarm.

The policy inertia was buttressed by high stakes politics. In September 1990, as the forces for Operation Desert Shield were beginning to deploy to the Middle East, President George H. W. Bush and the Democratic congressional leaders held a budget summit at Andrews Air Force Base in Washington, DC. The summit produced an agreement that was intended to curtail the federal deficit, but also may have cost President Bush his re-election bid. It was the moment that he broke his Republican convention promise of no new taxes. The deal created a firewall between defense and non-defense spending. Cuts in defense spending could not be used to finance increased social spending. Instead specific tax increases had to be enacted to fund additional spending. This removed the incentive for Congress to cut defense further to pick up money for favored social programs. President Bush knowingly alienated members of his own political party to protect the defense budget from raiding.[8]

The Democrats' successful focus on economic fears in the subsequent presidential election caused the Republicans to re-emphasize security fears during the Clinton years in an effort to reclaim political primacy. There was no agreement about where the threat lay or how big it actually was – it might be a resurgent Russia, a rising China, Middle East turmoil or the chaos in failed states – but the defense budget had protectors. The defense budget drawdown after the Cold War was the most gradual and modest of all the drawdowns since the end of World War II.[9] And with stabilizing resources and no clear security challenge, the American military apparently sought comfort in familiar priorities and old doctrines.

President George W. Bush's Secretary of Defense Donald Rumsfeld found himself fighting quixotically against those familiar priorities and doctrines. Appointed in 2001 with a mission to make transformation happen, Rumsfeld quickly ran afoul of entrenched Pentagon interests. His efforts to cut costs and change old ways of doing business were ill received by a military that had been assured "help is on the way" – a theme used by Republicans in the 2000 election to underscore their concern about America's defenses. Enraging the military brass and doing a poor job of representing the administration on Capitol Hill, many political observers were speculating before September 11th that Rumsfeld was on the way out. Even after September 11th saw a skyrocketing of Rumsfeld's popularity, his plans for transformation made little headway. Large infusions of cash were directed at traditional priorities, and outside of a few fights with the Army, little change was evident to outsiders. The second inter-war period ended dramatically with the attacks on the World Trade Center and the Pentagon, but the significance of the war they generated for transformation is obscure. The public's patience with Rumsfeld ran out well before the President sought and obtained his resignation five years into the war. Transformation's run has been longer, but apparently no more successful.[10]

The second inter-war period is over. Rumsfeld's replacement, Robert M. Gates, has admonished the services, and the Air Force in particular, for not supporting sufficiently the forces deployed to Iraq and Afghanistan. He complained about the slow deployment of UAVs and the attempts to continue the production run of the stealthy F-22 interceptor. There were also reports about dissatisfaction with the size and pace of the acquisition of mine resistant vehicles by the Army and Marine Corps. Coming late in the Bush Administration, however, it is unclear whether or not the dissatisfaction turns into doctrinal redirection and organizational transformation. More likely, it will be the administration replacing Bush's that will make the choices that will force either more attention to counter-insurgency operations or signal the start of an Iraq syndrome where interventions abroad are rare for US forces and none tempt an occupation. Both inter-war periods began with a fleeting belief that the world has just experienced a war to end all wars and took about two decades to complete. The issue is whether or not they both produced innovation.

Innovation or not innovation

Understanding the roots of the American military's interest in RMA technologies does not tell us much about the impact that the adoption of these technologies have had on the American military. Important questions remain: has the American military changed the way it fights? Does technological change mandate doctrinal and organizational changes? We have referred to the use of RMA technologies, but has there been innovation simply because the military sees better, communicates better, and can hit certain types of targets more precisely than it once did?

These puzzles compose our central inquiry: has there been real innovation during the second inter-war period, and if so, what are its causes? Pursuing this inquiry requires some definitions of key terms. We define the *Revolution in Military Affairs (RMA)* to be the set of technologies that enhances the ability of a military to locate a relatively greater set of targets and destroy them with precision. That is, anything we can see we can hit, and anything we can hit we can kill – and we can see a great many things. These technologies protect the users, for example with stealth technology that allows attackers to approach targets undetected. They also permit users to confine the destruction of targets, as with laser-guided weapons, so as to avoid the killing of innocents.

Innovation we define to mean significant changes in organizational tasks and rewards in the service of a major change in organizational output.[11] Significant change is defined relative to the organization, but in the case of militaries it is clear that change, if it is to be called innovation, requires upheaval in organizational hierarchy and doctrine. A faster jetfighter changes neither who runs the Air Force nor how the Air Force fights wars. In contrast, the adoption of submarine-launched ballistic missiles by the Navy gave that service a dominant role in strategic warfare, reduced greatly the interest in the US acquiring better strategic bombers, and afforded submariners the opportunity, if only fleeting, to be the senior leaders of the Navy. It also changed the focus of submarine operations from tactical to strategic.[12]

Innovation therefore is not simply the adoption of new technology. It is the forceful abandonment of the old. As Joseph Schumpeter, the great economist, put it so effectively, innovation is creative destruction. Something new is accepted at the cost of something old. With the advent of ballistic missiles and advanced air defense systems, bombers and their pilots fell from grace in the Air Force. Within the Navy, submariners pushed aside aviators and surface warfare specialists for the top slots. Some gain in status and power from innovation while others lose. That explains the opposition to innovation. The politics of implementing significant change is the political struggle over organizational or system rewards. Military innovation changes the way the military fights and who succeeds within the hierarchy. Bomber pilots long resisted the importance of fighter escorts during World War II and beyond, believing that they could carry the fight to the enemy without assistance. They had reason to be unhappy with technological change, as it was part of a decades long process that displaced them in favor of fighter generals. It is not irrational to oppose innovation, as it destroys opportunities as well as creates them.

Transformation in our view is the specifically organizational half of innovation. We define it as the organizational implementation of significant change. Transformation requires changes in military tasks, priorities, and rewards. The RMA gave the military new technology. Transformation, if it were to occur, would give the military new leaders, doctrine, and dominant coalitions. It should be visible both in process and in consequence. The losers in the struggle for innovation should seek to block or temper change that occurs. The effects of transformation should be visible in budgets and promotion lists.

Americans in particular tend to view innovation as always a good thing. To say that someone is innovative is to praise that person highly. Those who oppose innovation are dismissed as idiots and Luddites. In the military there is the cartoonish figure of the battleship admiral or the horse-loving colonel who failed to embrace respectively aircraft and the gasoline engine. But not all innovations are worthy of adulation. Consider the combat glider and the atomic cannon, both likely more dangerous to their users than to their enemies. Moreover, not all opponents of change deserve ridicule. In both Britain and the United States there were those who saw strategic bombing as neither moral nor effective. Despite the claims to the contrary, strategic bombing did not allow victory without the need to suffer horrendous ground combat casualties and it did focus death and destruction upon civilians. Debate and resistance are required to separate the truly good from the merely new among innovations.

Innovation theory

With innovation so highly valued in American society it is not surprising that there is much attention paid to organizational designs that supposedly promote innovation. Some say it is large, rich, monopolistic organizations that are the most innovative while others believe it is the small and vulnerable among organizations that are most likely to change their output.[13] On the one side there is the need to have sufficient slack resources to develop and test new ideas. On the other there is the argument that organizations have to want to change. Competitors can jeopardize the survival of an organization.

James Q. Wilson has theorized that organizations that are decentralized and diverse in terms of their skill base are more likely to generate ideas for innovation but are exactly the kinds of organizations that have trouble adopting new ideas. Diversity leads to counter offers, the proposal of innovations that favor your team or sub-unit against rivals and conflict over opportunities, the battle for the choice jurisdiction, and thus to difficulty in gaining agreement on the path to take. More centralized, less diverse organizations, in his view, generate fewer ideas, but can more easily make decisions.[14]

The American armed services are, of course, large, rich *and* diverse organizations. They have hierarchy *and* decentralization. In Wilson's scheme it is indeterminable how such organizations will behave. They will have lots of ideas about change and lots of difficulty choosing among them. Another perspective on organizational innovation helps us understand when the services might be innovative. Sanford L. Weiner, comparing federal research agencies, has argued that innovation requires both autonomy and uncertainty. Recall that we noted that some thought innovation needed organizations with slack, the excess resources of large autonomous, monopolistic organizations, while others thought that it was lean, fearful, competitive organizations that would innovate. For Weiner it is a mix. Organizations need to feel some vulnerability, some fear of the future, but they also need some autonomy, the time to meet a challenge. The services can fit these requirements as they are semi-independent organizations that are at least at times rivalrous, competing for the same roles and missions. Under these conditions the services have the resources – time and money – to innovate, but also the impetus to innovate, the fear of losing future status to a rival.

Taken together, the standard political science theories on military innovation help reinforce this judgment about at least some of the conditions needed for innovation and, more especially, how innovation will occur. One, closely associated with Stephen Rosen at Harvard, argues that military innovation is a gradual, generational process within militaries. Advocates of change find protectors, experiment doctrinally, and slowly climb the promotional ladder, contending with rivals for control over the direction of a military service.[15] Another, associated with Barry Posen of MIT, argues that significant doctrinal changes in militaries require the rare and direct intervention of senior civilians who, concerned about the capability of the nation to survive a likely war with a rival nation, move to force their military to adopt a winning but suppressed doctrine.[16] The third, much favored by this team, argues that military innovation, like innovation in the broader society, comes from competition among organizational rivals for resources and standing, who overcome the costs of implementing change through their concern about the greater costs of losing to a competitor.[17] The competing rivals in the case of the RMA are the armed services of the US military, which are spurred to innovate by competition among them over budget shares and status.

Innovation in practice

The case studies that follow build on this experience and the theories that we have outlined. Each chapter pursues the question of transformation by tracing the history of its service or topic throughout the second inter-war period. The chapters emphasize the organizational and doctrinal responses to the technologies of the RMA, in an effort to detect possible innovation and deduce the causes of its presence or absence.

To achieve this goal, actual organizational and technical changes are compared to two ideas of what transformation "ought" to have looked like. Both standards are suggested by the history at the beginning of this chapter; they are the most obvious innovative ways the military might have interpreted the technological and international signals it was receiving during the second inter-war period.

The first is the RMA ideal of how the relevant service could use technology to transform itself for high-intensity warfare – what we call "conventional transformation." This standard is the one the Soviets initially feared at the end of the Cold War: a reconnaissance strike complex that would allow Americans to find and kill so quickly it would completely negate their advantage in numbers. In a post-Cold War world, the effect would be an incredibly fast-moving and lethal force that could shred an opposing army with relatively few casualties. This interpretation at how the RMA might have been interpreted by the services is a worthy one, as it was the purpose for which most of the technologies in question were designed, and one much discussed in military circles. Evidence for meeting the conventional standard varies from case to case, but usually includes three criteria: the speed and manner with which a service incorporated RMA technologies; some cut in force structure that reflects the new efficiencies gained from the technology; and a change in doctrine reflecting a new concept of operations.

The second comparison is to the changes the service might have made to adapt to its environment – to the wars it fought and the enemies it seemed likely to fight, or "environmental transformation." Such a standard is justified by the type of operations that were actually carried out in the post-Cold War period. These were counter-insurgency, peace-enforcement, peacekeeping, and humanitarian relief – in short "small wars." Given the number of times America eventually engaged in these activities, there was a strong political signal that the military might have heard and adapted too. After all, conventional war doctrine and technology can often be cumbersome, clunky, and out of place in such operations, making this an area ripe for some creative destruction. The criteria for meeting this standard also vary but can include: major doctrinal changes; changes in training and emphasis on new operations; and of course relevant cuts and adjustments in force structure. Each author brings a different perspective to the task. In the end, little transformation was discovered, and the authors offer various explanations for this outcome.

To ascertain the causes of innovation, we employed the predictions of the three theories of military innovation mentioned above. Rosen's theory of intra-service implies a longer, generational timescale for innovation perhaps only reaching culmination in recent years. We therefore asked the authors to look for evidence from the Cold War period, if relevant, of innovative origins for service activities today. Rosen's theory also seems to argue for a competition among potential innovators for the support of senior officers within their own service. Each chapter thus checks for the relevant pet projects of senior officers, and whether they came to dominate other possibilities in the past, and contemporary operations in the present.

Posen's theory of civilian intervention has one very simple prediction: the heavy hands of civilians should be found in military organizations. We should observe them reaching down into the depths of a service to find dissenting officers pushing innovative ideas, promoting them, funding them, and backing their efforts. This should be more evident as signals from the international environment become more intense, and policymakers see activities they want to engage in abroad.

Finally, the theory of inter-service competition should show services developing doctrine and force structure for the same roles and missions. We should find them concerned about the lessons of recent operations, and interested in the opportunities the other services seem to be seizing. We should see them searching their environment looking for big changes that may jeopardize their future status among the services and thus future budgets. They should be sensitive to the demands of political masters and trends in military technology and doctrine.

Our authors found little innovation, noting in most instances that the factors we described above were absent. Their findings are previewed below. Our analysis of the American military's experience with the RMA in the years since the Cold War ended is discussed in the concluding chapter.

To understand the transformational effects or lack thereof stemming from technological change, it is necessary to have some background on the technologies in question. David K. Burbach, Brendan R. Green, and Benjamin H. Friedman begin in Chapter 2 with a discussion of the four major technologies of the RMA: sensors, communications networks, precision weapons, and stealth. After discussing the development history of each technology, they outline the major capabilities that appeared in the late Cold War period and some of their potential. The chapter concludes with some skepticism about how far the RMA vision can be taken.

The heart of the volume begins with the recent history of the different services, including Special Operations. Colin Jackson offers his assessment of the US Army's performance during the second inter-war period in Chapter 3. As Jackson's title suggests, the Army's attitude was initially characterized by a conservative approach to the RMA technologies. The Force 21 program produced some incremental changes, such as the much vaunted blue-force tracking capability, but by and large the Army was more concerned with fighting on familiar budget battlefields than either the conventional battlefields of the future or in the missions offered it by the interventionism of the 1990s. The Kosovo conflict flipped this attitude on its head: under the leadership of General Eric Shinseki, the army became interested in revolutionary visions offering rapid deployment and lightning mobility in place of heavy armor and firepower. It thereafter pursued this transformation assiduously, if not entirely successfully, despite technological limits and the conflicting priorities of two wars.

Next, in Chapter 4, Benjamin H. Friedman reviews the progress of the Navy in assimilating new technologies. Friedman notes that the Navy did adapt somewhat to the post-Cold War environment and that it evolved significantly through the incremental addition of new technologies to the fleet, especially in the areas of fleet defense and precision. However, it did not much alter its force structure. He therefore finds that transformation did not occur – neither by the standards of the political science literature on innovation nor by those of the more popular literature on the RMA. Friedman also notes that there is a marked and growing gap between the Navy's rhetoric on strategy and the reality of its procurement agenda. It is a Navy with a large appetite for technology, but little for organizational change.

Sanford L. Weiner follows with an analysis of the Air Force in Chapter 5. He tracks the Air Force's interest in precision weapons and strategic bombing from their origins, and gives an overview of Air Force performance during the major air operations of the 1990s. Weiner rates the Air Force's performance during much of the second inter-war period as good and its assimilation of technology as rapid. But he also judges that little organizational change has occurred, and in-depth examination of the development of Air Operations Centers and UAVs reveals that the Air Forces technophilic reputation is not entirely deserved. In the end, Weiner argues that transformation of the Air Force has yet to occur.

In Chapter 6, Austin Long examines the Marine Corps. Long concludes that the Marine Corps gestured at all kinds of changes during the second inter-war period but actually made very few, and almost none related to the technologies of the RMA. His central insight is the existence of two separate communities in the Marines – one dedicated to maneuver warfare in the style of the German army, and one more focused on small wars and counter-insurgency operations. The latter, more responsive to the operations American forces undertook and contemplated after the Cold War, focused on urban operations; the former concentrated on procurement of the V-22 helicopter and developing maneuver doctrine. Though the Corps underwent several changes, Long shows they were mostly rhetorical – the Marines cannot be said to have transformed.

The evaluation of the services concludes with a look at Special Operations Forces (SOF) in Chapter 7. Jackson and Long argue that SOF have achieved genuine transformation in the second inter-war period. Large bureaucratic changes have been made, with US Special Operations Command now a "fifth service" in terms of its ability to procure and operate independently from its parent services. The RMA technologies have made SOF an extremely useful part of the reconnaissance-strike complex, as they can deploy sensors and networks effectively to generate intelligence and targeting information. The emergence of counter-terrorism as the US military's principle task has made SOF the leading service in the militaries' largest global campaign, changing SOF's role from supporting to supported. However, Jackson and Long caution that this transformation has not been all to the good – the traditional missions related to counter-insurgency that SOF was created for are being abandoned. In their place, a preferred mission set centered on direct military action and intelligence gathering raids is being instituted based on an untested theory of victory. In the one service where transformation has definitely happened, it may not be an occasion for applause.

Chapters 8 and 9 look at the effect of RMA technologies on important supporting elements of the defense department – the Office of the Secretary of Defense (OSD) and the defense industrial base. In Chapter 8, Robert Tomes provides background on how the RMA vision evolved within the OSD in the period after Vietnam. Tomes shows that many of the ideas and technologies dubbed "revolutionary" in the 1990s actually dated to the 1970s, when Secretary of Defense Harold Brown and Undersecretary of Defense for Research and Engineering William Perry conceived of an "offset strategy" to address strategic and operational challenges posed by numerically superior Soviet conventional forces. Brown, Perry, and a handful of service leaders funded early experimentation with the technologies that would lead many to hail a "new American way of war" by Desert Storm. Tomes traces the intellectual trajectory and basic continuity in ideas about the RMA during the three decades following Vietnam, showing that despite tremendous technological change, much of defense transformation has actually been "old wine in new bottles."

Finally, in Chapter 9, Eugene Gholz examines the RMA's effects on the defense industry. He compares the history of the defense industry in the second inter-war period with a standard of how transformation ought to have affected it – the proliferation of information technologies should have driven defense contracts toward new technology firms who could produce low-cost systems based on optimizing platforms for a single task. These platforms would then combine to be more effective than the sum of their parts through the RMA technologies of networks linking sensors and precision shooters, providing the systems base for the revolutionary vision. In fact, the opposite occurred: large defense contractors outcompeted smaller technology-based firms in integrating the RMA technologies into military procurement plans. Gholz cautions that this failure to transform was not necessarily bad. In many ways the older defense firms had the necessary knowledge and skills to become excellent providers of the new technology.

At the same time, they also developed some serious acquisition pathologies, like the Lead Systems Integrator approach, putting all the government's eggs in one basket. Transformation in the defense industry did not occur, it appears, and the effects of technological changes are mixed.

In sum, we found that by the metric of conventional transformation, only three services approached partial credit: the Army, Navy, and SOF. No service produced environmental transformation. Despite large technical change the RMA has made relatively little organizational impact. The completed chapters reveal this plainly: the services did explore some alternatives, but seemed uninterested in making very large changes in their priorities. The interest in major platforms and major wars remained fixed. Secretary Rumsfeld's call for transformation had been sounded, but no one rallied to the cause. The wars that came along were not anticipated doctrinally even though they had many been preceded by many similar provocations. The contrast with the years between World War I and II, when important military innovations were conceived and adopted, and those between the Cold War and the current wars, when change was much touted but largely unattempted, is dramatic.

Notes

1 The Marine Corps suffered 24,511 deaths from all causes in World War II, while the Army Air Corps had 54,700 killed in action. Thousands more airmen died in training accidents. See *American War and Military Operations Casualties: Lists and Statistics*, CRS Report Order Code RL32492; and Air Force statistics online at www.usaaf.net/.

2 For a brief summary, see Andrew Krepinevich, *The Military Technical Revolution: A Preliminary Assessment* (Washington, DC: CSBA, 2002), pp. 5–7.

3 Daniel Wirls, *Buildup: The Politics of Defense in the Reagan Era* (Ithaca, NY: Cornell University Press, 1992).

4 The literature on the RMA is voluminous. For a representative sample, see Eliot Cohen, "A Revolution in Warfare," *Foreign Affairs* (March/April 1996); John Arquilla and David Ronfeldt, eds., *In Athena's Camp: Preparing for Conflict in the Information Age* (Santa Monica, CA: RAND, 1997); and Admiral William Owens with Edward Offley, *Lifting the Fog of War* (New York, NY: Farrar, Straus, and Giroux, 2000).

5 Lorna Jaffe, *The Development of the Base Force 1989–1992* (Washington, DC: Joint History Office, Office of the Chairman of the Joint Chiefs of Staff, 1993).

6 Ibid. It is unclear whether Powell was trying to save the forces from Congress' cuts or using this story as a way to get the services and the DoD to sign off on the cuts.

7 William W. Kaufmann, *Assessing the Base Force: How Much is too Much?* (Washington, DC: Brookings Institute Press, 1992).

8 David Mervin, *George Bush and the Guardianship Presidency* (New York, NY: St. Martin's, 1996), p. 134. Special caps were placed on defense, foreign aid, and other discretionary spending mandating that any spending increases within a capped area had to be met with other spending cuts in the same capped area, or paid for with tax increases; this is within the broader context of a budget deal that cut spending and raised taxes across the board.

9 Harvey Sapolsky and Eugene Gholz, "The Restructuring of America's Defense Industry," *International Security* (Winter 1999–2000), p. 15, chart.

10 For information on Rumsfeld's tenure and transformation outcomes, see the previously cited sources and the following contemporary accounts: "Comments by

Bush and Rumsfeld," *New York Times*, December 29, 2000, p. A18; "B1 Blundering," *Washington Post*, July 9, 2001; Rowan Scarborough, *Rumsfeld's War* (Washington, DC: Regnery Publishing, 2004); and the 2001 Quadrennial Defense Review.

11 James Q. Wilson, "Innovation in Organization: Notes towards a Theory," in J. O. Thompson, ed., *Approaches to Organizational Design* (Pittsburgh, PA: University of Pittsburgh Press, 1971), pp. 195–218.

12 Owen. R. Cote, Jr., *The Politics of Innovative Military Doctrine* (Dissertation, Cambridge, MA: Massachusetts Institute of Technology, 1996); and Harvey Sapolsky, *The Polaris System Development: Bureaucratic and Programmatic Success in Government* (Cambridge, MA: Harvard University Press, 1972).

13 Schumpeter asserted both, first favoring small organizations and then big ones.

14 James Q. Wilson, "Notes toward a Theory of Innovation," in James D. Thompson, ed., *Approaches to Organizational Design* (Pittsburgh, PA: Pittsburgh University Press, 1971).

15 Stephen Rosen, *Winning The Next War: Innovation and the Modern Military* (Ithaca, NY: Cornell University Press, 1991).

16 Barry Posen, *The Sources of Military Doctrine: France, Britain, and Germany Between the World Wars* (Ithaca, NY: Cornell University Press, 1984).

17 Cote, *The Politics of Innovative Military Doctrine*; Harvey Sapolsky, "On the Theory of Military Innovation," *Breakthroughs*, Vol. IX, No. 1 (Spring 2000); "The Interservice Competition Solution," *Joint Force Quarterly* (Spring 1996).

2 The technology of the revolution in military affairs

David T. Burbach, Brendan Rittenhouse Green, and Benjamin H. Friedman

This volume seeks to understand the technological changes that did – and did not – occur in the US military during the second interwar period. A major contention of several authors is that while technological change was rapid and impressive during the 1990s, doctrinal and especially organizational change was lacking. Advances in technologies were grafted onto the services' traditional missions, enhancing the combat power of existing platforms and ways of doing business. Before proceeding to a discussion of individual service histories, however, some background on the technologies that are said to present the opportunity for a revolution in military affairs is needed. Only by having a firm sense of the technologies involved is it possible to recognize some of their practical limitations, to analyze the manner in which the services actually employed them, and to evaluate their suitability to the current strategic environment.

Although there are many ways of describing the Revolution in Military Affairs (RMA), commonly noted technological components of the RMA include advances in information, stealth, and precision. Early visionaries of the revolution, both in the United States and in the Soviet Union, predicted not just new technologies, but new ways of using information to link together platforms to heighten combat capability. In the late 1970s, Undersecretary of Defense for Research and Engineering William Perry famously summarized this vision of warfare, noting that he wanted the United States

> to be able to see all high value targets on the battlefield at any time; to be able to make a direct hit on any targets we can see; and to be able to destroy any target we can hit.[1]

Later predictions would often feature diagrams showing a vast array of existing and imagined platforms all capable of communicating instantly and seamlessly to identify and destroy time-sensitive targets at long ranges.

Perry's statement provides a way to categorize the technologies developed. In this chapter, we focus on the ability to "see" everything on the battlefield through advances in sensors; the ability to "hit" everything that is seen, by exploiting networks between sensors and shooters, as well as stealthy delivery; and the ability to "kill" everything that is hit, through advances in precision weapons and submunitions. For each of these technologies – sensors, networks, stealth, and precision – we provide basic definitions, a sense of the technology's place in the RMA framework, a brief history of the most important innovations, and an explanation of where its development stands today. Some technologies, like stealth, demand a focus on individual services, while other technologies, like networking, cross service boundaries. These inquiries reveal the potential and the limitations of the technologies vis-à-vis the revolutionary vision and provide context for subsequent chapters' discussion of organizational politics and doctrinal decisions.

Sensors

Sensors are the eyes of the RMA. Sensors locate, track, and distinguish objects of interest over wide battle areas. Three types of sensors are worthy of special attention: radars, visual and infrared imagers, and sensors of signals intelligence (SIGINT includes electronics intelligence, or ELINT, and communications intelligence, or COMINT).

The past three decades have seen great advances across these areas, contributing to much greater clarity of "sight" on the battlefield. The number and types of platforms on which these sensors can be mounted expanded during the same timeframe, increasing the purview of the military's battlefield "vision." The historical arc and present capabilities of the three types of sensors are explored below.

Radar emits radio waves and receives their return as they bounce off objects in their path. After a target has been illuminated with radio waves, the signal return can be processed to determine the range, speed, direction, and altitude of the target in question. Radar's advantages include the ability to "see" beyond visual range in all types of visibility and weather conditions. In some contexts, radar-guided munitions can deliver ordinance with precision. Ideally, powerful and/or well placed radars would be sufficient to illuminate an entire battlefield, revealing the location and disposition of all potential targets in/on the air, ground, and sea; they would also do so regardless of weather conditions, terrain, and state of the target. The evolution of radar technology shows much progress toward this RMA ideal.

The military use of radar began in World War II. Radar was instrumental in air defense for both Britain and Germany, as it could locate incoming bombers and direct intercepting fighters. It was also useful for offensive air operations, when radar returns often presented the only reliable way to identify the cities that bombers were supposed to attack. On the sea, naval forces used radar to improve gunnery, direct fleet air defense, and to detect submarines operating on the surface.[2] Military use of radar had grown significantly by the Vietnam War, but still had important limitiations. Tactical and technical hurdles hurt air-to-air combat performance in Vietnam, for example. American radar-guided missiles – most prominently the AIM-7 Sparrow – required a minimum distance in order to activate, and were often fired short of such distances. They also required constant illumination of the target in order to complete an engagement, and radars could not yet "look down" to targets below the emitter because radar returns from the ground would wash out moving targets in "clutter." This limited the potential for radar in air-to-air combat. At the same time, powerful enemy ground-based radars with ground-controlled fighter interceptors served the same role they did in World War II, giving the defenders far better situational awareness than the strike packages that they faced. As a result, Vietnam-era engagements were reminiscent of the World War II/Korea era: fighters were usually killed by cannon before they were aware of the enemy.[3]

Five technical advances gave radar new applications after Vietnam. First, the development of pulse Doppler radar erased the ground clutter problem by filtering out radio returns from objects that were moving relative to the ground clutter. An airplane traveling several hundred miles an hour could be distinguished from the ground below it, and therefore targeted – the development of "look-down/shoot-down" capability. Second, miniaturization put radars in missiles themselves, allowing them to illuminate and track targets on their own – the capacity to "fire and forget." Third, the move from mechanically scanned to solid-state phased array radars meant that a single radar could illuminate one or more specific targets while continuing to scan the skies for others – "track while scanning" capability. Fourth, better analysis of the radio returns from different types of engines (known as NCTR) and radio signals that could interrogate both friendly and hostile aircraft for their identity (identify friend-or-foe, or IFF) allowed fighter aircraft to identify their targets before they could see them, meaning that air-to-air engagements could occur "beyond visual range." Finally, airborne warning and control system (AWACS) platforms were developed. They deployed sophisticated and powerful air surveillance radars. Flying behind the battlefield, AWACS platforms can "see" for several hundred kilometers and disseminate their battlespace picture to other platforms (see pp. 19–25 on networks below) for situational awareness and targeting information. By the time of the Gulf War, radar was approaching the RMA vision of being able to see everything on an aerial battlefield and kill it before visual contact.[4]

Recent radar technology has started to advance this vision with regard to ground targeting. The development of ground moving target indicator (GMTI) and synthetic aperture (SAR) radars (usually now different modes on the same sensor) has made it possible to track vehicles while they are moving at night and in all weather conditions, and image and precisely locate them when they are still. GMTI is a refined version of the pulse Doppler innovation that created look-down/shoot-down capacity and is able to track even slowly moving ground targets like tanks or other armored vehicles. Multiple networked GMTI radars tracking the same target from different angles can precisely locate its coordinates for GPS-guided munitions. SAR, on the other hand, takes very high resolution images of the ground. This means that stationary vehicles can be detected, accurately located, and often positively identified in relation to surrounding terrain, with obvious implications for targeting. The presence of both modes on the same sensor means that mobile targets can be continuously tracked.[5]

GMTI/SAR has been mounted on airborne platforms to allow coverage of large areas. The Joint Surveillance and Target Attack Radar System (JSTARS) is an AWACS-like aircraft that flies in friendly airspace while monitoring the ground battlespace for many hundreds of kilometers alongside it. Its powerful GMTI/SAR radar can locate and track a plethora of ground targets, and JSTARS can make this information available throughout the theater. Similarly, the development of High Altitude Long Endurance (HALE) unmanned aerial vehicles (UAVs) with GMTI/SAR capability has further extended the range and persistence of ground-intelligence. The Global Hawk UAV, in particular, can fly higher than JSTARS, cover a larger range with its radar, and can do so for days rather than hours before landing.[6]

Optical and imaging infrared (IIR) sensors provide different advantages. These are passive sensors that do not rely on illuminating the observed area to generate information, which means that the sensor does not emit radiation and give away its presence on the battlefield. Optical and IIR sensors are also less susceptible to jamming and often provide higher resolution information for targeting. Like radar, IIR sensors can be used at night, but they are not capable of seeing through clouds, and they generally have shorter detection ranges.

Optical sensors are simply photographic or video cameras mounted to a platform and trained at a target. In a manner familiar to anyone who has used a camera, they detect visible light to create images, these days at increasingly higher resolutions and from greater distances. IIR sensors rely on the part of the electromagnetic spectrum just below visible light. They absorb and analyze the infrared radiation emitted by all objects, relying on the tiny temperature differences between them to create an image that can be used to target from the sensor's platform or to cue other platforms. Because IIR sensors do not rely on visible light, they can work with or without daylight and through some types of battlefield obscurants that block optical sensors.

Optical and IIR sensors evolved in response to the needs of national intelligence, particularly in monitoring Soviet strategic programs and deployments. Both sensors rely on three major platforms – airplanes, satellites, and UAVs. Early in the Cold War, the only way to obtain vital information about Soviet missile bases and test sites deep inside Russia was to fly spy aircraft with cameras over the country, a problematic solution that resulted in the 1960 downing of a U-2 piloted by Gary Powers. This helped push the intelligence and military communities toward optical imagery satellites. The first versions of these satellites, the KH-8 and KH-9, took photographs from space and then ejected fully exposed film canisters back to Earth, where they were retrieved and processed. The much more recent KH-11 is an electro-optical system that translates photons into electronic signals that allow near-real-time download of collected images.[7]

IR imagery came into currency in a similar way, in response to the need to track Soviet missile launches. Operational since 1970 and capable of worldwide coverage, the US Defense Support Program (DSP) geosynchronous satellites can detect the infrared signature given off by missiles' rocket boosters and relay the location, missile type, and azimuth to a network of fixed and mobile ground stations. Those ground stations, in turn, can use Link-16 (see p. 23, below) to communicate the information to ships, tactical aircraft, and surveillance aircraft like AWACS and the Navy's E-2C Hawkeye. Although originally designed for detection of ballistic missile launches, they can also detect some cruise missile launches and jet afterburners.[8] Today, the intelligence community operates several types of satellites which between them cover the Earth in visible, infrared, and radar imagers.

Conventional aircraft and UAVs also contribute to the optical/IIR capability of the military. The Air Force still flies U-2s, while the Navy has adapted the P-3 ASW platform to collect imagery over land. UAVs like Global Hawk, Predator, and many smaller UAV have become premier imagery gathering platforms. All these aircraft carry electro-optical, IIR, and radar systems with the capacity for near-real-time download of intelligence and extremely high resolution of images.[9]

Optical and IIR sensors have important tactical uses. Imagery databases make it possible to get very precise coordinates for objects in a particular image. By knowing the location and attitude of the platform that took an image and comparing that image to known terrain features, it is possible to determine very precise coordinates for a GPS-guided weapon. In any event, knowing roughly where something is enables one to put a laser designator on the target if the weather is clear. Radar returns do not have this advantage unless one also has a radar-guided weapon. IIR sensors have broadened the scope of the targeting that can be done at night. The addition of targeting pods with IIR sensors to tactical aircraft has greatly increased the relevance of IIR sensors to precision targeting. The AGM-65 Maverick missile was originally developed as a TV-guided weapon but, in its newest form, it has an IIR sensor that allows for accurate targeting at night. The LANTIRN pod and its more recent competitors, now on essentially all US tactical aircraft, use their IIR sensors to detect potential targets using thermal imagery and then hand them off to Maverick missiles upon acquisition. The thermal imaging of LANTIRN also allows the use of laser-guided weapons, since the system also has a laser designator that can be cued by the pod.[10]

Signals intelligence or SIGINT sensors are a third category of sensors important to the RMA. SIGINT refers generally to the collection and analysis of electromagnetic signals and is divided into Electronics Intelligence, or ELINT, and Communications Intelligence, or COMINT. COMINT is the interception of enemy communications signals, conveyed by radio, satellite, telephone, or other methods. ELINT collects other electronic signals, including radars, missile telemetry, electronic interrogating equipment, and others. After sensors have received SIGINT, it is decoded or interpreted as necessary and can be used for targeting under the right circumstances.

SIGINT has several advantages for these purposes. Like radar, it is an all-weather-capable form of intelligence. Unlike radar, SIGINT sensors are passive and thus do not give away the location of the sensor. Unlike either radar or optical/IIR sensors, SIGINT can offer longer detection ranges because propagation is not necessary limited to line-of-sight paths (i.e. radio signals can bend around terrain or be extended by atmospheric scattering). The major disadvantage of SIGINT over radar is that establishing a target location – both direction and range – requires multiple bearings from several widely separated angles.

Precision targeting with SIGINT has only recently become possible using the method of Time Distance of Arrival (TDOA). TDOA uses mathematical formulas to exploit the differences in time that it takes a given signal to arrive at physically separate but networked sensor locations. With three separate time-of-arrival measurements, a unique point on the globe can be identified whence the signal must have come.[11] Measured with enough accuracy, such a point is sufficient to program a GPS-guided weapon. In addition, precision targeting with COMINT can be simpler when the content of the intercepted communication specifies the target's location or provides information from which its location can be deduced. This is often said to be relevant to low technology/high profile targets, such as terrorists.[12]

ELINT took off in World War II, where it was used primarily against air defense radar. ELINT sensors provided a rough location of the radar so that photo-recon missions could discover their exact location for later attack. They could also gain intelligence about the radar that enabled countermeasures to be deployed: frequency, pulse characteristics, and signal strength could be determined and used for jamming. Though most sensors were originally located on aircraft, by the early 1960s, satellites with ELINT capability had been launched, primarily to gather data on Soviet radars. By the mid-1970s, COMINT receivers were in place on geosynchronous satellites, and space-based capabilities grew steadily as the Cold War progressed. While important strategically, orbiting sensors are less than ideal for tactical targeting because the great distances involved mean that even TDOA can only give approximate locations of emitters.[13]

The growing number of platforms with SIGINT/ELINT capability is eliminating the problem of having too few receivers for highly accurate triangulation. Ships, airplanes, UAVs, and ground stations have made SIGINT TDOA possible in many situations. Airframes are the major contributors to SIGINT targeting today, such as the Air Force's RC-135 RIVET JOINT, the Navy's EP-3 Aries, and the Army's RC-12 Guardrail. These platforms can locate, analyze, and identify communications and radar signals in a very short amount of time. When networked with other SIGINT sensors, they provide a precise location for the emitter. UAVs are also acquiring ELINT and COMINT capacity, and as they come online in greater and greater numbers, the scale of SIGINT surveillance will increase, as will the ability to exploit TDOA networks. This is particularly relevant to the high-value targets and other time-sensitive operations relevant to counterterrorism.[14]

Networks

Advances in computers and electronics underpin all of the RMA technologies.[15] Improvements in sensor capabilities, for example, required not only improvements in the actual receivers but the development of digital signal processing technologies to extract information out of the received signals. Precision depends upon miniaturized, automated systems to steer munitions toward designated targets. Stealth aircraft could not fly without computerized flight control systems.

While information technology has contributed to the whole range of military capabilities, the most radical development has been networking. Militaries have long sought to see more clearly, direct fire more accurately, and avoid detection – telescopes, rifled gun barrels, and camouflage are just a few examples – but it is networking technologies that allow the shared knowledge that advocates say is transforming warfare. At its core, the vision of the RMA is that whatever is known by any element in networked military forces can be known immediately by every element, thus "lifting the fog of war" (in Admiral Owens' formulation) to create what the military calls dominant battlespace knowledge.[16] This capability not only supports William Perry's vision of being able to destroy any targets that can be seen, but goes beyond that to allow radically decentralized operations – hopefully allowing much greater speed and agility – by providing all levels of the military with the full battlespace picture.[17]

The history of the development of communications and computer networks covers a particularly wide range of discoveries and applications. The basic technologies are employed in a vast range of military and civilian systems. If stealth represents one extreme of a highly specific technology developed and exploited by the military, networking represents the other: a suite of generic and highly adaptable technologies that today are almost entirely developed commercially and then used by the military.

The emergence of military telegraph systems in the nineteenth century ushered in the modern era of command and control, in which it became possible for generals to have real-time information and control of forces outside their immediate vicinity. That ability took a much greater leap in the first half of the twentieth century as radio technology matured. While World War II technology was primitive by modern standards, subject to significant range and reliability problems and only supporting voice or slow teletype communication, it still allowed great advances in military forces' ability to coordinate operations and react to changing situations.

During the early Cold War period, Defense Department needs played a leading role in the development of computer technology and communications networks.[18] After World War II, some observers saw the future potential of networked computers, sensors, and precision weapons. In 1948, for example, Jay Forrester of Harvard suggested that within fifteen years digital computers could take over much of air traffic control and the command and control of air defense networks.[19] And in fact, it was air defense that brought the first harbinger of automated, networked military operations.

American officials were deeply worried by Soviet bombers in the 1950s, and set out to create an air defense network to protect the country. Given the vast area to cover and the speeds of jet bombers, the system had to be highly integrated and automated. What emerged in the late 1950s was the Semi-Automatic Ground Environment or SAGE system – a network of twenty-three control centers linking dozens of radars, anti-aircraft missile sites, and airbases. The core of each center was a pair of MIT-designed, IBM-built computers that collated radar data, generated situational displays, and computed real-time interception and guidance information.[20] The wartime effectiveness of the fragile system was questionable and ICBMs rendered it obsolete, but SAGE brought breakthrough technologies like magnetic core memories, modems, and CRT displays. More generally, it demonstrated the potential of real-time automation of military operations.[21]

Throughout most of the Cold War, the priority in developing command and control systems was on requirements for nuclear war. In some ways, nuclear command and control needs were simple: operations would have involved a relatively small number of platforms executing pre-determined strikes against known and relatively immobile targets like silos, airbases, and defense industrial plants. On the other hand, communications would have to be conducted in a hostile environment (including jamming, electromagnetic pulse, and the destruction of many communications nodes) and require extreme reliability. Little communication among different tactical units would be needed during nuclear operations, and so relatively little attention was paid to interoperability or networking.

By the end of the Eisenhower era, some DoD leaders questioned this approach, and in 1960 Secretary Thomas Gates created the Defense Communications Agency to promote integration and centralization of C3I networks.[22] This push for improved and centralized C3I increased during the Kennedy administration. The administration's "flexible response" policy and counterforce doctrine called for a wider range of nuclear strike options and survivable command and control. The Cuban Missile Crisis also taught Kennedy that civilian leaders needed more direct means of communication with deployed tactical forces during crises.[23] One outcome was the creation of the National Military Command System, later renamed the World-Wide Military Command and Control System (WMCCS). Developed in the mid-1960s, WWMCCS was to tie together existing communications systems (including non-military locations like the White House, CIA, and State Department).[24] The results fell far short of what was hoped for due to service recalcitrance (and the lack of authority for OSD or DCA to overrule them).[25] Nevertheless, it did represent the first major step toward integration and interoperability of communications systems.

Another advance of the 1960s was the development of communications satellites. The Defense Department aimed to launch the world's first communications satellite, and although AT&T took that prize, by the late 1960s, DCA followed with the Defense Satellite Communication System.[26] Individual services also developed their own satellite systems, notably the Navy's FLTSATCOM. The services also leased capability from commercial satellite providers.

The 1960s and early 1970s also saw the development of three fundamental technologies that enabled military applications and the civilian IT revolution: modern computer networking, interactive human–computer interfaces, and microprocessors. Computer networking developed most directly from military research. To support reliable communications during a nuclear conflict, the Defense Advanced Research Projects Agency (DARPA), the RAND Corporation, and other organizations developed technologies that allowed distributed networks of computers to communicate with each other reliably and efficiently, even over disrupted communications paths.[27] Though intended to ensure post-nuclear command and control, these technologies were a prerequisite for tactical data networks (and for "the Internet").

Computer interfaces also became more suited to real-time tactical operations. Early computers had cumbersome input–output technologies, such as punched cards and printouts, and were optimized for "batch" processing – running single programs one after another. The late 1960s saw the development of software that allowed interactive use of computers by many users simultaneously, as well as now-familiar input–output devices like the video display and the "mouse." The Intel corporation also developed microprocessor integrated circuits – "chips" sufficiently small and robust to allow capable computers to be fielded in aircraft, ships, and tactical vehicles.

While the principal focus in this era was on strategic nuclear systems, the war in Vietnam drove some communications innovations. The Army fielded tropospheric scatter radios capable of relaying numerous voice communications channels over extended ranges, for example.[28] A major Vietnam-era development in network warfare was the so-called "McNamara Line" or IGLOO WHITE system designed to stop traffic on the Ho Chi Minh Trail. At its peak after 1970, this network included thousands of acoustic, seismic, magnetic, and other sensors to detect North Vietnamese traffic, with aircraft relaying signals to a computerized control station in Thailand where detections were correlated with available strike aircraft, assigning them targets.[29] Based on such advances, in October 1969 General Westmoreland gave a speech outlining his vision of the "revolution" in warfighting that could result from an Army "built around an integrated area control system that exploits the advanced technology of communications, sensors, fire direction, and automatic data processing."[30]

The 1970s brought advances in networking that would be recognizable to observers of the contemporary US military. It became possible to deploy robust digital networks that could handle high-speed voice and data traffic using common protocols.[31] In addition to basic technological improvements in existing networks, this period saw significant service investment in platforms and units devoted largely or solely to communications. Attempts to integrate service efforts increased, but with mixed success. The services retained budgetary authority over C3I, so they were free to effectively ignore OSD recommendations. Not surprisingly, they resisted investment in equipment that would give civilians greater control over tactical engagements.[32] They also had little interest in improving interoperability and so projects like the OSD-initiated Joint Tactical Communications Project, known as TRI-TAC, fared poorly.[33] In contrast, the less-ambitious JCS effort called JINTACCS (program for Joint Interoperability of Tactical Command and Control Systems) accepted that the services would develop different systems but aimed to increase interoperability through "jointly acceptable protocols and message formats between different digital data system in such areas as air operations, intelligence, fire support, operational control, and amphibious warfare."[34] JINTACCS did not involve new hardware development, but rather adoption of new formats and procedures to standardize messages and make them more machine-readable. By the 1980s, it had made some limited contributions along these lines.[35]

Even more significant was the Joint Tactical Information Distribution System (JTIDS), also known as "Link-16". Launched in the mid-1970s, and first fielded in the early 1980s, JTIDS was the most successful civilian-pushed networking effort of this period. Originating in the Office of the Deputy Secretary of Defense for Research and Engineering, JTIDS was a new common architecture for data communications. Based on the networking principles developed in the 1960s, JTIDS allowed distributed users to exchange data over a digital packet-switched network with much greater speed and security than previous systems (such as the 1960s "Link 11" systems).[36] Because this system had no headquarters, only mobile nodes, it was less vulnerable to enemy attacks. Service cooperation remained problematic: Navy participation in JTIDS only came in 1985 after political pressure in the wake of communication failures during the Grenada operation.[37] Systems like JTIDS were critical to disseminating data from new tactical surveillance platforms like the Navy's E-2C Hawkeye, Air Force AWACS, JSTARS, and RIVET JOINT.

The Army also invested in its communication technology. Despite improvements implemented in Vietnam, most Army radios still dated to the 1950s and were insecure, bulky, and not able to handle digital data. The Army therefore invested in two new communications systems: SINCGARS (Single Channel Ground/Airborne Radio System), a new family of tactical radios able to "hop" frequencies to defeat jamming, and Mobile Subscriber Equipment (MSE).[38] MSE was the most important of these for future networking, in that it was a "tactical internet": a digital, packet-switched wireless network that allowed peer-to-peer communications and supported voice, data, and fax.[39] MSE would be the foundation of a variety of Army tactical data sharing and command systems, allowing automation of many routine aspects of operational planning and providing commanders with greater situational awareness.[40]

During the 1980s, many earlier service networking initiatives came to fruition, including several related to satellites. The DSCS constellation continued to expand, as did leasing channels on commercial satellites. Nevertheless, priorities continued to emphasize nuclear warfighting. The premier program of the 1980s was the MILSTAR satellite, which was to have "heroic" survivability measures and exceptional encryption and anti-jamming capabilities, at the cost of very low bandwidth – i.e. sufficient for a President to signal "bomb Moscow tomorrow" to surviving nuclear forces but not to support exchange of real-time data between users in a major conventional conflict.[41]

The 1990s brought three key changes. First, the end of the Cold War shifted the focus from strategic nuclear war to conventional operations, particularly regional conflicts versus nations like Iraq or North Korea. Second, in the wake of the 1986 Goldwater-Nichols Act, the ability of the Joint Staff and OSD to exert central control on defense programs grew considerably. Finally, by the 1990s, the locus of technology development had shifted overwhelmingly to the private sector. The explosive growth in civilian hardware and software made it possible for the Defense Department to focus on application and integration rather than invention.[42] Four developments were especially important: the maturation of microcomputers, the availability of wired and wireless networks, the emergence of generic client–server interfaces like the "world-wide web," and the development of interactive database and information retrieval systems.

First, during this era, microcomputer technology reached the point where capable systems could be fielded in many vehicles, command posts, or personal devices.[43] Second, fixed-line and wireless networking became ubiquitous in civilian settings, as did cellular phones, leading to explosive growth in data transmission speeds as well as a wide range of hardware and software solutions. While adoption of off-the-shelf systems was not always possible due to requirements for security and jam-resistance, these developments did provide key building blocks.[44] The third area of commercial development was the world-wide web and associated software tools that allowed many types of content to be delivered from any server and displayed meaningfully by a single "browser" program on any type of end user terminal.[45] A related development was database and transaction processing software that connected those web servers to large information stores. The net impact was that single, general-purpose terminals (like laptops or vehicle-mounted systems) could support many different applications (in contrast to old "stovepiped" systems), and that software developers could work far more efficiently – coding only the details of specific applications and allowing generic browsers and servers to handle communications and displays.

These developments together provided the technology base to fulfill the vision of the early RMA proponents: a world where every platform and soldier is connected in real-time to world-wide shared information resources. Bringing that into reality, though, requires far more than technological building blocks: equipment must be selected and fielded, military applications and databases must be created, etc. That effort continues to occupy the Pentagon.

Since the late 1990s, the Office of the Assistant Secretary of Defense for Networks and Information Integration, along with the Defense Information Systems Agency, has pushed a two-pronged approach to integrated networking. The first prong seeks to invest in new networking technologies, such as "new communication satellites, next-generation interoperable radios, a new ground-based communication network with significantly expanded bandwidth, and services and applications to manage and protect the network and help users locate, post, and share information."[46] The second prong seeks to integrate existing networking capabilities with the new technologies to create the Global Information Grid (GIG). The goal is to establish a secure, reliable network that any user can contribute to and receive from, regardless of service, platform, time, or location.[47] Key acquisitions underway to build the GIG network capability include a new constellation of communications satellites, new tactical radio systems, and bandwidth expansion of the DoD infrastructure.[48] Despite this ambitious agenda and the tens of billions of dollars that have been spent on GIG since the late 1990s, key aspects of the project remain problematic – the satellite program has not reached the hoped for level of technical readiness, for example,[49] and other hardware and software projects also face delays.[50]

Even if networked warfare does not live up to the expectations of its advocates, there is no doubt that modern communications and computing technologies have had dramatic impacts on military capabilities. These technologies provide the linkages and rapid information processing that make possible some sort of order out of the mass of data from new sensors and weapons platforms. In the early Cold War, much of the challenge was in developing fundamental new technologies, and in its first two decades many of the IT industry's breakthroughs trace back to military R&D programs. Today, commercial systems provide an abundance of building blocks for military network designers. The key challenges are developing military-specific adaptations and making both the technological and organizational changes needed to achieve interoperability among a multitude of systems. While significant progress has been made compared to previous decades, major challenges still loom in achieving the integrated and instantaneous awareness desired by transformation proponents.

Precision

Inherent in the concept of the RMA is the notion that targets can be identified, located, and attacked with great accuracy. Where for centuries munitions were unguided shells following ballistic paths to a target, the technical advances of the last several decades have allowed the creation of precision-guided munitions (PGMs). PGMs constitute a broad category of munitions delivery systems that use active guidance technologies to steer themselves to targets with a higher degree of accuracy than unguided munitions.[51] They are the "fists" of the RMA, providing the ability to destroy any target that can be located. Common examples of PGMs include cruise missiles, and air-to-ground guided bombs, and missiles that home on radar/radio emitters. Contemporary guidance technologies include lasers, GPS satellite signals, radar or infrared homing, even acoustic guidance in the case of some anti-armor munitions. This summary focuses on air-to-ground precision weapons.

The history of guided bomb development in the US military is the story of incremental and evolutionary technological and tactical development, punctuated by critical technological breakthroughs and illuminating combat experiences. The earliest aerial bombardment tactics, introduced in World War I, entailed aircraft flying together in tight formation, with only the lead aircraft using a bombsight to pinpoint the target. The other aircraft released their weapons upon receiving a signal from the lead aircraft. Such tactics were unable to reliably hit small targets, but the only other solution given World War I technology was to conduct bombing sorties at low altitudes, where planes would be at risk from anti-aircraft fire. This combination of tactical necessities and technological limitations created a niche for the first precision weapons.[52]

World War II provided the first true test of American airpower, with mixed results. In the aftermath of the war, the United States Strategic Bombing Surveys (USSBS) concluded that airpower, though a decisive factor in the Allied victory, had many serious limitations: the munitions were too small, very few bombs hit their targets, air defenses and radar increased aircraft losses and decreased the accuracy of the munitions, and inclement weather was a dramatically underestimated constraint on munitions accuracy.[53] Although bomber squadrons flying in formation and dropping unguided munitions did most of the bombing carried out by US forces in World War II, the Army Air Force (AAF) did experiment with guided bombs on a limited basis, included glide bombs and bombs guided by radio control.[54]

The end of the war brought a significant drawdown of forces and budgets, leaving little money available for projects outside of the development and delivery of nuclear weapons.[55] Vietnam brought a renewed interest in precision weapons. When air-to-ground operations began in Southeast Asia, the Air Force's Tactical Air Command soon made use of a Navy-designed bomb, the Walleye. The Walleye was one of the first "launch and leave" systems; its electro-optical homing system could automatically guide the 829-pound bomb to a predetermined target.[56] However, Air Force pilots found fault with the Walleye's targeting system, which had a tendency to target clutter within its sensor field rather than the desired target.[57]

As a replacement for the Walleye system, in 1967, the Air Force embarked on Project Paveway, to develop laser-guided bombs (LGBs). Aircrews and senior Air Force officers were so pleased with initial combat test results that they uniformly adopted the new weapons technology over its Navy counterparts.[58] Toward the end of the war, LGBs previewed the bright future for guided bombs. In one attack in 1972, eight aircraft equipped with sixteen LGBs attacked and destroyed five bridges. Inflicting the same damage with "dumb" bombs would likely have required more than 2,000 sorties.[59]

Many airpower analysts point to the 1991 Gulf War as a watershed conflict for precision engagement. Although 90 percent of bombs dropped in that effort were unguided, it is estimated that the 7 percent of munitions that were laser-guided did 75 percent of the damage inflicted by airstrikes. That performance, coupled with widely-publicized footage of LGBs striking buildings, bridges, vehicles, etc., greatly intensified the effort to shift from unguided bombs to precision munitions. Laser-guidance technology suffers from inherent limitations, however. The pilot must maintain line-of-sight visibility of the target from weapons release until impact, and laser designation will not work through clouds or smoke. In the course of the Gulf War, one out of every three laser-guided attacks was called off due to inclement weather conditions.[60] Cruise missiles were also extensively employed in the Gulf, but they were less accurate and cost $1 million each.

It was the frustrations of partially effective precision munitions in the Gulf War that were the impetus for an all-weather, cost-effective technology. By that time, a solution had appeared in the form of the Global Positioning System, which allowed munitions to determine their precise geolocation in any weather conditions, over any type of terrain.

Before the advent of GPS, guided weapons either had to "home" on targets illuminated by radar or laser designators, as discussed above, or used inertial guidance systems – gyroscopes and accelerometers – to track their motion relative to their known launch point.[61] Although gyros are commonly used to measure the attitude of airplanes and missiles (i.e. to know what "straight and level" is), inertial guidance lacks the accuracy to direct weapons to within a few meters of a target. High-precision inertial systems are very expensive to manufacture, and are somewhat bulky and fragile. While they were worth including on long-range nuclear missiles, inertial guidance systems were not a useful solution to guiding tactical munitions.

GPS works on the principle of "time-difference-of-arrival" analysis that was discussed earlier in the section on sensor technology.[62] In principle, a set of transmitters at known locations broadcast continuously, and a receiver picks up signals from three or more of them. Each broadcast communicates the current time at the transmitter's location, and the receiver then compares the time of broadcast to the time of arrival. That travel time gives the distance to the transmitter – signals always move at the speed of light – and with distances to multiple transmitters, triangulation will yield a precise location. In practice, GPS is somewhat more complicated. The transmitters are orbiting the Earth, and so their position is constantly changing. Also, there is no way for a receiver to have its own internal, independent time reference. These problems are solved by having the signal include each satellite's orbital path, so that the receiver can calculate where it would be at a given time, and by transmitting a long, known code sequence – the receiver knows that if "ABCCBA" was transmitted, that transmission would have happened at precisely time X.[63]

By the early 1960s, the military had seen the potential of space-based navigation on the TDOA principal, and the Navy, Air Force, and Army all began work on global navigational systems for their own use.[64] Most notable were the Navy's Transit satellites, which tested key technologies and provided limited navigational capabilities. In 1973, OSD forced the competing programs to be consolidated into an Air Force led joint program that combined the innovations made to that point by the Navy and USAF on their own. The basic technologies for satellites and receivers were mostly worked out by the late 1970s. Deployment during the 1980s was slower than expected. Despite the defense budget increases under Reagan during the early 1980s the services cut funding for GPS, using the funds to accelerate weapons procurement; the space shuttle *Challenger* disaster also slowed launches.[65] By 1991, a functional constellation of satellites was in place, and the system showed its worth in helping U.S. forces to navigate the featureless desert during the Gulf War.

Seeing the opportunity presented by GPS – and the limitations of then-current laser-guided bombs – in May 1991, Air Force Chief of Staff General Merrill McPeak wrote, "We need to lay down a requirement for an all-weather precision-guided munition." This initiative led to the joint direct attack munitions (JDAM) program, overseen by Air Force Major General Joseph Ralston.[66] The JDAM program was tasked with providing an all-weather, precision-guided munition that could be carried by both Air Force and Navy tactical aircraft. The solution, developed at Eglin Air Force Base, was to add a guidance kit to standard unguided bombs that used both GPS guidance and a backup inertial guidance system.[67] After successful initial tests, McDonnell Douglas and Martin Marietta (later Lockheed Martin) won contracts to develop competing designs for the initial phase of the JDAM program.[68] Reflecting the new priority given to precision, in July 1995, OSD agreed to accelerate the initial production decision on the JDAM by six months and the full production decision by fifteen months. As a result, in October 1995, the Air Force awarded the $62.6 million contract for the engineering and manufacturing development of the JDAM to McDonnell Douglas,[69] and actual production began by 1998.[70]

In 1999, the air war against Serbia demanded precision-guided munitions, but laser-guided bombs could not succeed in the cloudy Balkan weather. General Ralston, Vice-Chairman of the Joint Chiefs of Staff, initiated the use of JDAMs for these missions, but because the system was still in low-rate initial production, the munitions were in short supply.[71] This shortage limited the number of precision missions flown during the conflict, but even so, in forty-five B-2 missions, 656 out of 682 JDAMs were dropped successfully and 80 percent of their targets were destroyed or damaged.[72] Following the successful use of the JDAM system in Kosovo, the Air Force contracted for 11,000 JDAMs with Boeing for $235.5 million (a unit cost of just over $21,000), in March 2001.[73]

The development of guided bombs has depended on the incorporation of new technologies and encouraged significant tactical changes. The earliest precision weapons could not be "guided" in any meaningful sense; they could only be launched on a predetermined course, and, conditions permitting, maintain that course with some moderate degree of accuracy. The addition of visual sensors and radio control allowed World War II and Korean War airmen to control and correct the trajectories of their weapons, providing considerably greater accuracy than unguided bombs. However, the necessity for line-of-sight targeting meant that bomber crews were exposed to ongoing anti-aircraft fire while they guided their munitions to the target.

The development of "launch and leave" weapons, guided by electro-optical systems or lasers, solved many of the human error problems associated with line-of-sight targeting. Furthermore, the unprecedented accuracy of laser-guided bombs developed during the Vietnam War allowed for the most significant tactical change up to that point: the dramatic reduction in the number of sorties needed to successfully destroy a given target, such that the one-aircraft-one-bomb-one-target concept became technologically feasible. However, LGBs had significant limitations of their own related to line-of-sight targeting and the need to maintain a laser on the target. The JDAM and other GPS-guided bombs seem to have solved most, if not all, of the technological problems that previously limited guided bombs. This technological leap has tactical rewards. Now a single bomber aircraft carrying multiple JDAMs can destroy multiple targets with previously unmatched accuracy. Long gone are the days of multiple heavy bombers flying in formation and dropping unguided bombs miles away from their targets.

Stealth

Stealth refers to technologies that make a target less visible to sensors. Stealth technology can be applied to any military asset that can be sensed and destroyed, including manned and unmanned aircraft, cruise missiles, armored vehicles, submarines, warships, and even satellites. The most relevant sensors that stealth avoids are active and passive radars, infrared, and sonar. Most often stealth refers to efforts to protect aircraft against detection by radar.

Stealth's place in the RMA vision is this: by denying an adversary the ability to detect and locate one's own assets, it increases one's own survivability, and especially the ability of lightly protected platforms operating alone or in small groups to conduct deep strikes. Stealth – if possessed only by one's own side – makes it possible to conduct the bold, small-group attacks envisioned for "net-centric warfare."

Stealth incorporates a variety of technologies working on different physical principles. In the case of active radar, stealth reduces electromagnetic energy reflected back to the source (or to a passive munition homing on the radar signal) from the target – reducing its radar cross-section. Stealth protects submarines by quieting them to avoid detection by sonar.[74] Infrared detection is prevented by limiting heat emission. Among other things, this limitation protects against heat-seeking missiles. Platforms avoid heat emission by shielding and cooling exhaust, by shielding the hottest parts of the engine, and by eschewing large, heat-producing afterburner engines. Ships can more easily screen hot parts of the engine and cool exhaust. However, they produce heat in various other ways that are hard to mask.[75]

Integrated air defense systems use radar as their primary sensor. They use radar for three things. They conduct surveillance of airspace to locate targets. They use it for fire control, establishing the target's position and bearing, identifying it, and guiding weapons to its vicinity. They generally also use radar to kill the target by detonating a warhead within its lethal range.[76]

There are several ways for aircraft to avoid detection by radar. One is to simply avoid the radar by going under, over, or between the areas covered by the adversary's radar systems. Aircraft can also avoid it by firing standoff weapons like cruise missiles before they come within detection range. They can evade weapons directed by radar by outrunning them or flying above their maximum altitude. Radars can be jammed by radiating signals at the radar transmitter to interfere with it or confused by dispersing chaff or radar-absorbing aerosols. All of these methods have limitations, and so the most recent and revolutionary innovation is to design aircraft that are inherently difficult to locate by radar – that is, making them stealthy.[77]

There are two principal methods of design that avoid detection via radar: absorption and deflection. Absorption is a straightforward concept: platforms can be built or coated with materials that absorb radar signals instead of reflecting them. Deflection takes advantage of the fact that most radar uses the same antenna for receiving and sending, designing the aircraft so that very little energy is reflected in exactly the direction it came from, instead bouncing it off in other directions. This is accomplished by avoiding right angles and large flat surfaces, and instead relying on irregular facets, curved surfaces or a large number of small flat surfaces.[78] Stealth imposes design limitations. Weapons stores and fuel tanks must be carried internally to avoid reflections off external fasteners, doors, and the like. Avoiding large flat surfaces means dispensing with conventional stabilizers and even employing inherently unstable designs like the B-2's flying wing. Keeping such planes straight and level requires constant adjustment by computerized flight control systems.

Radar came into its own during World War II, and militaries immediately desired ways to shield aircraft from radar. Such technologies were unknown at the time however. There were some attempts, mostly unsuccessful, by the US Army Air Corps and the MIT Radiation Laboratory to develop radar-absorbing coatings. There was little knowledge of how airframe designs could lessen radar detection. Instead, aircraft dropped chaff – then thin strips of aluminum foil the length of half the wavelength of the enemy radar – to reflect enemy radar and mask bombers. Fighter escorts protected bombers from enemy interceptors keyed by radar detection of bombing parties. Speed and the sheer numbers of attackers also helped overcome defenses.

The Germans experimented more directly with stealth. They had some success in masking submarines from radar detection at sea by covering their snorkels in a rubbery material. They also began developing a twinjet fighter late in the war called Horton Ho IX, which used wood in its design to absorb radar. They never deployed it.[79]

The end of World War II quickly brought the Cold War and independence to the Air Force. Emboldened by its role in Japan's surrender, the Air Force saw its job as winning wars by directly attacking enemy nations and destroying their will to fight, especially through nuclear weapons. Missiles had not matured to carry atomic weapons to the Soviet Union, so the job fell to bombers. Bombers had to get through to drop their ordnance, and that required avoiding detection by Soviet radar. At the start of the Cold War, the principal means of doing this was to fly fast and high enough to avoid the fighters and anti-aircraft missiles cued by the radar – detection was accepted as inevitable. The Air Force also planned to use electronic countermeasures such as jamming and chaff to protect bombers.

There were some early attempts to reduce the radar signature of American aircraft. The arrival of jet engines at the end of the 1940s overcame one obstacle: propellers were nearly impossible to mask. Other technical hurdles were also high, particularly the fact that conventional metallic aircraft had large flat surfaces conducive to radar reflection. Building aircraft out of non-metallic materials available at the time was not an option; none had the requisite strength-to-weight, flexibility, and other characteristics needed in an airframe.

During the 1950s, engineers began to conceive of ways to design aircraft to overcome radar – principally through the use of curved surfaces. Once it became clear that the Soviets could track the U-2 spy aircraft, designers sought to reduce its radar signature, with limited success. One early attempt to design an aircraft with a low radar cross-section was the CIA's A12 reconnaissance aircraft, which evolved into the SR-71 and entered service in 1963. The wings were blended into the engine casing and the fuselage, and new radar-absorbing paints were used, among other measures. Still the aircraft, which flew at about 100,000 feet, evaded danger mostly by speed and altitude.[80]

Until the 1970s, the Air Force and the Navy were not highly interested in stealthy aircraft. This was a function of the technical limits discussed above and the fact that the Air Force – at the time dominated by the bomber-centric Strategic Air Command (SAC) – preferred bombers that avoided surface-to-air missiles with height and speed. That approach soon ran into limitations. Soviet surface-to-air missiles became steadily deadlier. By the late 1950s, the Air Force envisioned that bombers would need to fly at more than Mach 3 above 70,000 feet – the design goal for the B-70. Even after the U-2 shoot down in 1960 proved the Soviet ability to target aircraft at high altitudes, Strategic Air Command's position was that bombers should stay high and rely on speed and electronic countermeasures to avoid SAMs.[81] The Vietnam War and Israel's experience in the 1973 October War showed that Soviet radars and SAM systems had become increasingly effective against US warplanes. Ultimately, Defense Secretary Robert McNamara canceled the B-70 program and closed the B-52 and B-58 bomber production lines. McNamara believed that bombers should hide from radar amid ground clutter and insisted that SAC's entire bomber inventory be outfitted for low altitude missions.

While the threat environment was making stealth more desirable, designing stealth aircraft was also becoming more feasible. The ability to predict radar cross-section (RCS) returns with models improved dramatically during this period. Meanwhile, the development of carbon fiber components and high-strength plastics offered more opportunity to design aircraft out of radar absorbing materials. In the past, these materials could only be considered as covers for radars. In the 1960s, their ability to bear heavy loads improved. Developers looking to make aircraft production less expensive by using these materials stumbled onto their radar-resistant properties when air traffic controllers had trouble seeing them with radar at airports. Radar, it was noted, passes through composites and other plastic siblings.[82]

The other technological change that enabled stealth aircraft was fly-by-wire flight controls. A central problem in producing stealthy aircraft was aerodynamic stability. Stealthy designs avoid features traditionally necessary to aircraft stability, like tails. Fly-by-wire is an aircraft flight control system that uses electrical interfaces for control of the aircraft, providing reactions that pilots alone cannot. Flight control computers interpret pilot commands, determining how best to manage each control surface to produce the intended response. This type of system was not unknown prior to the 1970s; the breakthrough was in creating the necessary redundancy and reliability to make the design safe at a reasonable price.[83]

Finally, in the 1970s, the Air Force had developed AWACS aircraft with radar that could look down and identify low-flying aircraft. Although the Soviets were never able to deploy such systems during the Cold War, at the time it was assumed that they soon would.[84] This development threatened to take away the safety offered by ground clutter, making stealth necessary to achieve penetration.

The first aircraft designed initially for stealth was the F-117A strike aircraft. Prototype development began under Lockheed's Advanced Development Projects or "Skunk Works," with funding from the Defense Advanced Research Projects Agency (DARPA) in 1973. DARPA at the time funded research programs at five major contractors to make headway in stealth.[85] This research became the "Have Blue" flying prototype. Lockheed won the contract, which grew into the F-117A program in 1978.[86] The program, secret for its first five years, ultimately produced fifty-nine aircraft that entered operation starting in 1983. Unlike later stealth designs that avoid flat surfaces, the F-117A's exterior consists of a series of small, flat, carefully oriented planes, called facets, which deflect radar in various directions. Heat emission from exhaust is shielded and cooled to limit the IR signature. The F-117A also extensively uses radar absorption material. It is built mostly out of a material termed Fibaloy, which consists of glass fibers embedded in plastic, and carbon fibers.[87] The F-117A has no afterburning engine and is subsonic.

The most famous stealth bomber, the B-2, began in 1977 under a contract to Northrop. The aircraft, whose existence was only acknowledged in 1980, revived the idea of a bomber as a flying wing, blending the wing entirely with the fuselage and cockpit to avoid sharp angles and flat surfaces. The exterior is all large-radius curved surfaces. The wing screens the exhaust, and the exhaust is cooled. The frame is titanium, but the exterior is made of graphite-based, radar-absorbing composites.[88] There is no tail. Nothing is carried externally. The B-2 has a terrain-following, low-altitude capability as well as a high-altitude profile – a change after the initial design that led to expensive avionics upgrades. The original plan was to buy 133, but due to the high cost and the Cold War's end, only twenty-one were procured.

The Navy's stealth aircraft in the 1980s was to be the A-12 II, the follow-on to the A-6, a carrier-based attack aircraft. The Navy selected McDonnell Douglas as the developer in 1988 and aimed to buy 620, along with the Marines' 238. The A-12 II had a flying wing design like the B-2, with internal bomb bays and all-curved surfaces. It was dubbed "the flying dorito." The aircraft was cancelled in 1991 due to the end of the Cold War and problems making the composite materials and radar-absorbing coating rugged enough to handle carrier landings and a sea environment. Among other problems, the A-12 II was much heavier than intended and far over budget.

During the 1980s, most American military aircraft began to acquire stealthy characteristics. Fighters are easier than bombers to make stealthy because of their small size. When the F-16A was upgraded in the early 1980s to the C version, it was given some features that made it stealthier from the front, including cockpit coating. The Navy's F-18 received more extensive stealth attributes upon being upgraded to the E version in the early 1990s, including radar-absorbent coatings, the hiding of intake valves, and some adjustments to materials on the wing. The

B-1 bomber, which was cancelled by Carter but revived by the Reagan administration as the B-1B, gained some stealth characteristics, including fuselage wing blending, increased use of composites, and changes to the engine intake materials.

Cruise missiles also got stealthier during the 1980s. These weapons include the Tomahawk, launched from submarines and surface ships, and the Air-Launched Cruise Missile (ALCM), later replaced by the Advanced Cruise Missile, for launch from aircraft. Cruise missiles had low radar cross-sections even when first tested in the 1970s. Upgrades reduced the radar signature of ALCMs by redesigning the nose, limiting hot exhaust, and streamlining the tail.[89] The ACM, now slated for replacement by the stealthy Joint Air-to-Surface Standoff Missile, has even more stealthy features, as do later versions of the Tomahawk.

When the Cold War ended, American contractors had extensive experience designing stealthy aircraft, and the services continued to aggressively pursue stealth. The F-35, or Joint Strike Fighter, was begun in 1996 with Lockheed as the lead contractor. The Navy and Marines plan to buy 480 each to replace the F-18, and the Air Force perhaps as many as 1,700 to replace the F-16. The F-35 is not all-aspect stealth in that it is not stealthy from the rear. It has a large internal weapons bay but can carry wingtip air-to-air missiles and external ordnance when stealth is not needed. The F-35 achieves a low radar signature by using advanced composite materials and a specially designed nozzle. The wings, fuselages, and canopy are sloped. The engine face is hidden and its inlet avoids leaving space between the duct and the fuselage to avoid reflecting radar energy.

The F-22 has been in development since the 1980s, when it was pegged as the Air Force's replacement for the F-15, the Air Force's premier air-to-air fighter. It has the radar cross-section of a marble, according to the Air Force, about the same as a B-2 bomber. Like the F-35, the wings, fuselage, and canopy avoid angles. The F-22's canopy has no frame, limiting radar returns from the windshield area. The engine intake duct is hidden. Weapons can be carried internally, although the load is relatively small. The F-22 uses faceting with blended facet boundaries. It uses few radar-absorbing materials (RAM), relying mostly on deflection. Exhaust nozzles use RAM to reduce radar and IR signatures, however, as do the wing edges. The F-22 also has a paint that reportedly reduces its IR and radar signatures.

Stealth features are also being incorporated into American UAVs. However, a debate is afoot as to how stealthy UAVs should be. More stealth means higher cost. If the point of UAVs is being cheap enough to be built in large numbers and used in risky operations, expensive stealth features would negate their advantage over piloted aircraft.[90]

Several factors limit what stealth can do for American airpower. Because fixed targets are often destroyed by cruise missiles and other aircraft, bombers must seek relevance by finding mobile targets. But this can require turning on radars to find the target, exposing the bomber to passive radar receivers. Also, even the stealthiest aircraft are occasionally detectable by surface radars at certain angles.[91] Additionally, low-frequency radars can identify most stealth aircraft. Though they cannot pinpoint location, they can guide SAMs that, despite inaccuracy, might drive aircraft to take defensive measures and devote large amounts of resources to their destruction or suppression. As a result, stealth bombers are still accompanied by jammers.

The stealth concept is also finding naval applications. Submarines have always sought to limit their noise to avoid sonar. Missile and aircraft range have made the Navy very interested in limiting the radar signature of its ships. Ships have a problem similar to aircraft in that they travel in a medium that makes them stand out. American warships can employ angles and absorbing materials to lessen their radar reflection, but they are huge relative to aircraft. This makes masking them far more difficult and limited in effect. Moreover, even if a ship manages to limit its own radar cross-section, it often carries objects like helicopters and small boats which have higher RCS, undermining the effort. American surface ships also have a broad array of antennae, usually on the mast for maximum range, which create a good radar target. Ships also tend to transmit radar and radio signals and to give off heat and noise.[92]

Still, the Navy wants to reduce some ships' radar cross-sections significantly. The aim is not to make ships invisible to radar, which is not feasible, but to reduce their radar cross-section enough that they might be confused with merchant vessels, and to allow chaff, decoys, jammers, and other methods to help defeat cruise missiles in their terminal phase. Moreover, lower radar cross-sections can allow a Navy ship to detect and destroy enemy ships and aircraft first, before they detect it.

The newest American destroyer, being built by Northrop Grumman and General Dynamics and dubbed the DDG-1000, is intended to have a reduced radar signature comparable to a merchant ship.[93] This is to be accomplished by a wood–plastic composite deck, a curved, inward sloping hull that avoids right angles, and less obtrusive missile launchers. The guns disappear into angular housings and helicopters are contained in the rear deck.[94] The hull is wave-piercing, meaning it cuts through rather than going over waves. This feature reduces the pitching and rolling that makes ships better radar targets. The end goal is a radar cross-section that is one-fiftieth of the current class of destroyers, the Arleigh Burkes. At a price of $3.6 billion per ship, naval analysts question whether the attempts to reduce radar signature justify such risky and expensive designs.[95]

The Navy's newest ship, the Littoral Combat Ship, is also meant to be stealthy. The theory is that its small size and ability to thwart radar will allow it to operate in range of surface guns. A trimaran hull design and use of composites are expected to reduce radar signature. Composites allow lighter weight, which makes ships faster, aiding speed and survivability. However, critics argue that the Littoral Combat Ship will still be detectable by a range of sensors, including radar, and so would be quite vulnerable should it operate in range of land-based sensors.

Conclusion

Several general points about these four categories of technology are relevant. They all are products of long development processes. No technology emerged as a bolt from the blue in the way that nuclear weapons did. Most trace their origins to World War II and have been developed over decades. What is new is their maturation and their combination into integrated systems.

The evolution of these capabilities was produced by political and organizational demand well as technological supply. Students of international relations and defense often debate the relative power of technological developments and policy decisions in shaping the international environment. For example, for proponents of the "offense–defense balance" theory of war that claim technological change itself can cause conflicts: when technology favors the offense, war is more likely. Their intellectual opponents believe that technologies are often developed to support policymakers' preferred military doctrines and that the causal weight lies in the politics of the nations that make wars and their rivalries.[96] RMA enthusiasts believe that technological change itself can radically alter military organizations and warfare. Others believe that technology emerges only when it serves the political goals of some organization or state.

The stories of these four capability areas lend weight to the argument that policy choices drive technology development. In all four cases, key advances were made after defense planners saw a need for capabilities, and the capabilities developed tended to reinforce rather than redefine the services' preferred methods of warfighting. In the case of stealth, for example, the big push for the technology came only after it became clear that the faster-higher approach to penetrating Soviet air defense was doomed to failure. Even then, if not for the Air Force's deep commitment to piloted strategic bombers, the US might have simply given up that leg of the triad and relied upon a large and reliable arsenal of land and submarine based missiles – as the Soviets did – rather than pursue the very expensive development of stealth technology. Precision-guided munitions development emerged from frustration over ineffective bombing campaigns in Vietnam, slowed for twenty years, then jumped ahead after the Gulf War.

The importance of policy preferences is even clearer when one considers that engineering breakthroughs alone cannot transform warfighting capabilities – institutional decisions must be made to procure and integrate systems based on those breakthroughs. Existing technology in the 1980s would have allowed the military to move more rapidly on fielding RMA technologies, for example. More laser-guided bombs could have been purchased; tactical radios and data links could have received higher priority. Certainly, the services could have made much more progress on networking communications systems, where in many respects the military lagged behind the commercial sector. The roadblocks to networking were as much organizational as technical.

Rather than being technologically determined, procurement and doctrinal decisions were consistent with national strategy and service preferences: priority went to strategic nuclear platforms and post-Armageddon command-and-control systems like the MILSTAR satellite. While some defense planners had imagined RMA-style capabilities decades earlier – Westmoreland in 1969 for example – the transformation of conventional combat capabilities did not become the overriding focus of the Pentagon until the early 1990s. The changed strategic and budgetary environment after the Cold War had as much to do with that shift as did the availability of new technologies.

This is not to say that scientific progress does not matter. No general could have simply ordered the B-2 bomber or a laptop computer into existence in 1950. The nuclear weapons story shows that new military capabilities can appear unasked for from the laboratory, but such cases are the exception. Fundamental levels of technology development shape the universe of possibilities, but the direction of R&D is shaped by existing priorities. Which experimental products eventually become fielded systems depends on institutional choice.

If strategies and doctrines lead militaries to pursue one out of a variety of technological paths they might choose, then what options were not chosen during the period since the Cold War? Other chapters in this volume explore each service's innovation or lack of it, but there are a few observations to make. Most important is that the technological base of the 1990s could have supported a variety of developments. As noted earlier, tremendously powerful and flexible general purpose computing technology came into being, with rapid advances in all manner of fronts coming from the commercial world. There was also rapid progress made in biotechnology, materials science, and other fields. Basic science and engineering could have allowed defense planners to take focused military R&D in a number of directions.

In some cases, solutions to needs observed in the 1990s would not have required much of a technology push: Somalia and Bosnia showed the need for a light, wheeled vehicle offering more protection from small arms and mines than the "Hummvee," yet none was developed until casualties mounted in Iraq. In terms of new areas of technology, a Defense Science Board report from 2006 recommended several items for priority development and, interestingly, many differed from traditional RMA focus areas.[97] The DSB emphasized technologies for understanding "human terrain," as well as information technologies for building up and maintaining long-term knowledge of human networks, fleeting or clandestine targets, etc. – most RMA visions have emphasized the "here and now" picture over correlating low-intensity activity throughout a protracted conflict. Finally, the services might have simply chosen not to put as many scarce dollars into R&D and deploying "transformed" combat systems and instead maintained larger low-tech force structures appropriate for stability operations, peacekeeping, and counterinsurgency.

The Cold War technological base combined with the computer revolution and other advances of the 1990s provided a long menu of capabilities from which the military services could choose to focus their development and procurement efforts. The rest of this volume will explore in detail the choices that each service made, the organizational and doctrinal changes that did or did not follow, and their relative degrees to which their internal evolution prepared them for their new external environment.

Notes

1 Perry cited in Richard Van Atta and Michael J. Lipptz, *DARPA's Role in Fostering and Emerging Revolution in Military Affairs* (Alexandria, VA: Institute for Defense Analyses, November 19, 2001), Assault Breaker chapter.
2 Excellent treatments of radar in World War II can be found in Louis Brown, *A Radar History of World War II* (Philadelphia, PA: Institute of Physics Publishing, 1999) and Henry Guerlac, *Radar in World War II* (Washington, DC: Tomash Publishers, 1987).
3 Marshall L. Michel III, *Clashes: Air Combat over North Vietnam* (Annapolis, MD: Naval Institute Press, 1997) is a thorough treatment of air combat in Vietnam, including the technical and tactical aspects of radar. See, *inter alia*, pp. 14–20, 285–290.
4 George Stimson, *Introduction to Airborne Radar, 2nd edn* (Mendham, NJ: Scitech Publishing, 1998) is a good introductory text covering most of these innovations, esp. pp. 317–322, 388–390, 548–549. See also Col. C. R. Anderegg, *Sierra Hotel: Flying Air Force Fighters in the Decade after Vietnam* (Washington, DC: Air Force History and Museums Program, 2001), pp. 9–11, 32–33; and P. Mason Carpenter, *Joint Operations in the Gulf War* (Maxwell Air Force Base, AL: Air University Press, 1995), pp. 11–18.
5 Stimson, *Airborne Radar*, pp. 403–424, 434–435.
6 Ibid., p. 549; for an exhaustive examination of the capabilities and development history of UAVs, see Thomas Earhardt, "Unmanned Aerial Vehicles in the United States Armed Services: A Comparative Study of Weapons System Innovation" (Dissertation, Baltimore, MD: Johns Hopkins, 2000).
7 Jeffrey Richelson, *The US Intelligence Community* (Boulder, CO: Westview Press, 1999), pp. 150–155.
8 Norman Friedman, *Seapower and Space* (Annapolis, MD: Naval Institute Press, 2000), Chapter 10, especially 242–243. For networks, see p. 358, fn. 18; for background on DSP satellites, see *Defense Support Program: Ground Station Upgrades Not Based on Validated Requirements*, GAO/NSIAD-93-148, May 1993.
9 Richelson, *The US Intelligence Community*, pp. 161–167.
10 Anderegg, *Sierra Hotel*, pp. 119–142.
11 This is an approximation – three measurements will yield two points, one of which will be in outer space. Accuracy issues in the measurement of the time implies that the more platforms, the merrier in order to reduce error in the location. Other methods of exploiting signal characteristics are also available, such as analysis of the signal's angle of arrival. TDOA is the most common and effective method. For a difficult but useful study, see Paul Chestnut, "Emitter Location Accuracy Using TDOA and Differential Doppler," in *IEEE Transaction on Aerospace and Electronic Systems*, Vol. AES-18, No. 2 (March 1982), pp. 21–28.
12 For example, see Harold Brown and James Schlesinger, "A Missile Strike Option We Need," *Washington Post*, May 22, 2006, for a possible use of such intelligence.
13 Richelson, *The US Intelligence Community*, pp. 185–190.
14 Ibid., pp. 191–197.
15 Adm. Bill Owens, *Lifting the Fog of War*, with Ed Offley (New York: Farrar, Straus, and Giroux, 2000), pp. 14–15, 99–100. For a critical review of claims about the role of information in the RMA, see Jeremy Shapiro, "Information and War: Is it a Revolution?" in Zalmay M. Khalilzad and John P. White, eds., *Strategic Appraisal: the Changing Role of Information Warfare* (Santa Monica, CA: RAND, 1999), pp. 113–153.
16 Owens, *Lifting the Fog of War*, pp. 138–143.

17 The Pentagon term for this vision is "network-centric warfare." Office of Force Transformation, US Department of Defense, *Implementation of Network-Centric Warfare*, 2005, Washington, DC. Online, available at: www.oft.osd.mil/library/library_files/document_387_NCW_Book_LowRes.pdf. Of course, that level of connectivity would also support micromanagement by generals or civilian leaders; moving to decentralized, non-hierarchical operations is a function of organizational change, not something that inevitably flows from the technology itself.

18 For a general history of the government/military role in the development of computers during World War II and the early Cold War, see Kenneth Flamm, *Creating the Computer: Government, Industry, and High Technology* (Washington, DC: Brookings, 1988).

19 Paul Edwards, "A History of Computers and Weapon Systems," in David Bellin and Gary Chapman, eds., *Computers in Battle: Will They Work?* (Boston, MA: Harcourt, Brace and Jovanovich, 1987), p. 54.

20 Robert Buderi, *The Invention That Changed the World* (New York, NY: Simon and Schuster, 1996), pp. 383–398.

21 Ibid., pp. 400–403. IBM's highly successful 700 series computers were directly derived from SAGE's FSQ-7 computers. SAGE work was crucial in establishing IBM's dominance over competitors like Burroughs and Sperry-Rand.

22 David E. Pearson, *The World-Wide Military Command and Control System: Evolution and Effectiveness* (Maxwell, AL: Air University Press, 2000), p. 19.

23 Friedman, *Seapower and Space*, p. 55.

24 Pearson, *The World-Wide Military Command and Control System*, pp. 52–61; see also Bruce Blair, *Strategic Command and Control: Redefining the Nuclear Threat* (Washington, DC: Brookings, 1985), pp. 53–56.

25 Pearson, *The World-Wide Military Command and Control System*, pp. 66–68. Pearson details the exceptionally poor performance of the system during the 1967 *Liberty* incident, due to both technical failures and command failures. Ibid., pp. 71–78.

26 Donald H. Martin, "History of U.S. Military Space Programs," *Crosslink* (Winter 2002). In the early 1960s the Pentagon's "Advent" program aimed to field a three-axis stabilized geosynchronous satellite. In practice, that satellite technology did not mature until the late 1970s.

27 For a general history, see Janet Abbate, *Inventing the Internet* (Cambridge, MA: MIT Press, 1999). On the relevant technologies, see Andrew Tannenbaum, *Computer Networks, 4th edn* (New York, NY: Prentice-Hall, 2002). The first step was the development of "packet switching" by Paul Baran of RAND in the early 1960s, a technology for distributed digital communications. At the same time, researchers connected with DARPA became interested in general methods for communication between computers, even if different models running different software. By 1970, a network of research computers called "ARPANET" was created – the progenitor of today's Internet. ARPANET had no operational military use, but the technologies and protocols developed for it were used in future military systems.

28 Maj. Gen. Thomas M. Rienzi, *Vietnam Studies: Communications-Electronics, 1962–1970* (Washington, DC: Department of the Army, 1972), Chapters 1, 6. Online, available at: www.army.mil/cmh/books/Vietnam/Comm-El/index.htm.

29 Paul Dickson, *The Electronic Battlefield* (Bloomington, IN: Indiana University Press, 1976), pp. 83–91; John T. Correl, "Igloo White," *Air Force Magazine*, Vol. 87, No. 11 (November 2004).

30 Dickson, *The Electronic Battlefield*, p. 222.

31 Robert R. Fossum and Vinton G. Cerf, "Communications Challenges for the 80s," *SIGNAL*, Vol. 34, No. 2 (October 1979), pp. 17–25. Reprinted in S. J. Andriole, ed., *High Technology Initiatives in C3I* (Washington, DC: AFCEA International Press, 1986).

32 Blair, *Strategic Command and Control*, pp. 57–60.

33 Kenneth Allard, *Command, Control, and the Common Defense*, revised edn (Washington, DC: National Defense University, 1996), pp. 204–205.
34 Ibid., pp. 204–207.
35 Ibid., p. 208.
36 John Williamson, ed., "Joint Tactical Information Distribution System," *Jane's Military Communications 2007–2008* (Surrey, United Kingdom: Jane's Information Group, 2007).
37 David Morrison, "Joint Acquisition of Major Weapons Likely to Remain an Elusive Strategy," *National Journal*, Vol. 18, No. 1 (January 4, 1986), p. 22.
38 SINCGARS ended up suffering significant technical and production delays, however, and was not actually fielded until the late 1980s. Jack Robertson, "Army Eyes Off-the-Shelf Radios for SINCGARS as ITT Lags," *Electronics News*, Vol. 32 (April 14, 1986), p. 66.
39 Allard, *Command, Control, and the Common Defense*, pp. 182–183.
40 William S. Pennypacker, "Automation: The Commander's Key to Victory in Airland Battle or Another Source of Friction?" (Leavenworth, KS: School of Advanced Military Studies, 1987), pp. 28–32.
41 Ben Ianotta, "Milstar: Mission Impossible," *New Scientist*, No. 1948 (October 22, 1994), p. 42. Even during the 1980s, MILSTAR was controversial, and with the end of the Cold War, the program faced enormous criticism. Eventually, the satellites were redesigned to offer somewhat greater bandwidth and more utility for tactical forces. They were finally launched in the late 1990s – but still provided much less capability than commercial satellites (pre-nuclear war, at least).
42 National Academy of Sciences, *Realizing the Potential of C4I: Fundamental Challenges* (Washington, DC: National Academy Press, 1999), pp. 199–204. As the Academy report notes, in some areas military systems do have requirements that require fundamental technologies not well-developed commercially (e.g. anti-jamming or radiation hardening). On the other hand, they noted, the Pentagon still had many untapped opportunities to exploit cost-effective off-the-shelf systems.
43 There has been no basic change in the underlying technologies, rather, steady improvement in processor speed, memory capabilities, and networking bandwidth. One notable improvement of the late 1990s and early 2000s, however, was the spread of affordable color liquid crystal displays, which are far smaller, less power-demanding, and more rugged than their CRT predecessors. For a discussion of the general increase in computer power with military implications, see Naval Studies Board, *Technology for the United States Navy and Marine Corps, 2000–2035, Vol. 2* (Washington, DC: National Academy Press, 1997), Chapter 2.
44 Increasingly, the military is able to use off-the-shelf systems. The US has deployed cellular phone networks in cities in Iraq and Afghanistan that allow soldiers to use their personal cell phones from the US, and Alcatel-Lucent recently began supplying the military with a rugged, rapidly deployable "cell-system-in-a-box" so that entire battlefields can quickly be provided cell service. Carolyn Marsan, "U.S. Troops Get On-the-Spot Mobile Access," *Network World*, March 15, 2007.
45 See Abbate, op. cit., also Tim Berners-Lee, *Weaving the Web* (New York, NY: Harper-Collins, 1999). The two basic innovations behind the "web" were hypertext markup language (HTML), a general protocol for content creators to use to describe how information should be displayed for a user, and the "browser" – software that receives content encoded in HTML and displays it on the user's machine per the content creator's direction. Since those initial developments, there have been many subsequent additions to allow multimedia and interactive content for example. The fundamental notion of separating content creation and content viewing remains, though: developers need only specify what to send and how it should look and leave the details to the recipient's browser. Users can leave the work to Microsoft's programmers and know that Internet Explorer will show them everything from weather data to email to *American Idol* reruns.

46 "Defense Acquisitions: the Global Information Grid and Challenges Facing Its Implementation," Government Accountability Office report to the House of Representatives, GAO-04–858 (July 2004).
47 "DOD Management Approach and Processes Not Well-Suited to Support Development of Global Information Grid," Government Accountability Office report to Congressional Committees, GAO-06–211 (January 2006), p. 7.
48 "Defense Acquisitions: the Global Information Grid and Challenges Facing Its Implementation," Government Accountability Office report to the House of Representatives, GAO-04–858 (July 2004), pp. 8–9.
49 "Defense Acquisitions: Assessment of Selected Major Weapons Programs," Government Accountability Office report to Congressional Committees, GAO-05–301 (March 2005), pp. 115–116.
50 "Defense Acquisitions: Resolving Development Risks in the Army's Networked Communications Capabilities Is Key to Fielding Future Force," Government Accountability Office report to the Chairman, Committee on Appropriations House of Representatives (Washington, DC: GAO-05–669, June 2005), p. 1; Paul L. Francis, "Defense Acquisitions: the Army's Future Combat Systems' Features, Risks, and Alternatives," testimony before the Subcommittee on Tactical Air and Land Forces, Committee on Armed Services, House of Representatives, GAO-04–635T (April 1, 2004), p. 1.
51 One might also include in the "precision revolution" some examples of highly accurate yet still ballistic munitions. For example, the M1 Abrams tank fires a ballistic shell, but the tank's fire control system is so accurate that a well-trained crew can often hit an opposing tank with just one shot.
52 Paul G. Gillespie, "Precision Guided Munitions: Constructing a Bomb More Potent than the A-bomb" (Dissertation, Lehigh University, 2002), pp. 28–29.
53 David R. Mets, "Long Search for a Surgical Strike: Precision Munitions and the Revolution in Military Affairs" (PhD, Maxwell AFB, AL: Air University Library/ Air University Press, 2001), pp. 8–9; Blackwelder, "The Long Road to Desert Storm and Beyond: The Development of Precision Guided Bombs" (MA thesis, Maxwell AFB, AL: Air University Press, 1993) pp. 1–40.
54 Blackwelder, "Long Road," pp. 5, 9; Kenneth P. Werrell, *Chasing the Silver Bullet: US Air Force Weapons Development from Vietnam to Desert Storm* (Washington, DC: Smithsonian Books, 2003), p. 139; Blackwelder, "Long Road," p. 10. The Germans also experimented with guided bombs, notably the "Fritz X" and "Henschel 293" radio-controlled anti-ship munitions.
55 Gillespie, "Precision-Guided Munitions," p. 65.
56 Ibid., pp. 142–143.
57 Anderegg, *Sierra Hotel: Flying Air Force Fighters in the Decade after Vietnam*, p. 122.
58 Werrell, *Chasing*, p. 149.
59 Ibid., p. 151.
60 George Cahlink, "Birth of a Bomb," *Government Executive*, Vol. 35, No. 11 (August 2003), pp. 21–22.
61 Donald McKenzie, *Inventing Accuracy: A Historical Sociology of Nuclear Missile Guidance* (Cambridge, MA: MIT Press, 1993). McKenzie provides a comprehensive overview of the workings and history of development of inertial guidance systems, particularly for long-range missiles.
62 The principle was also used in the older, land-based LORAN system. LORAN was important for maritime navigation and was sometimes used in aircraft, but did not have the accuracy, global coverage, speed, or digital automation needed for tactical munition guidance.
63 Elliot Kapan and Chistopher Haggerty, *Understanding GPS: Principles and Applications, 2nd edn* (Norwood, MA: Artech House, 2006), Chapter 2.

64 Scott Pace, Gerald Frost, Irving Lachow, David R. Frelinger, Donna Fossum, Don Wassem, and Monica M. Pinto, *The Global Positioning System: Assessing National Policies* (Santa Monica, CA: RAND, 1995), p. 238.
65 Ibid., p. 243. GPS was zeroed out entirely in 1980–1982, but was refunded at the insistence of OSD.
66 Cahlink, "Birth," p. 22.
67 Ibid., pp. 21–23. The inertial system would provide a backup in case GPS reception was lost during the bomb's descent.
68 "McDonnell Douglas, Martin Marietta Win JDAM Contracts," *Defense Daily*, Vol. 183, No. 9 (April 13, 1994).
69 "McDonnell Douglas Wins JDAM Development Contract," *Defense Daily*, Vol. 189, No. 8 (October 12, 1995).
70 Tom Breen, "First JDAM Guided Launch Was Successful, Officials Say," *Defense Daily*, Vol. 193, No. 16 (October 23, 1996); Tom Breen, "GPS-Aided JDAM Test Is Unsuccessful, Air Force Says," *Defense Daily*, Vol. 193, No. 24 (November 4, 1996); Tom Breen, "GPS-Aided JDAM Test at Eglin Is Successful," *Defense Daily*, Vol. 193, No. 41 (December 2, 1996); General Accounting Office, "Joint Direct Attack Munition: Low-Rate Initial Production Decision," p. 1.
71 Cahlink, "Birth," p. 27.
72 Robert Wall, "Pentagon Details Flaws in Kosovo Weapons," *Aviation Week & Space Technology*, Vol. 152, No. 8 (February 21, 2000).
73 "Boeing Nets $235 Million JDAM Full-Rate Production Contract," *Defense Daily*, Vol. 210, No. 1 (April 2, 2001).
74 George Lindsey, *The Tactical and Strategic Significance of Stealth Technology* (Quebec City: Centre Quebecois de Relations Internationales, 1989), pp. 9–10.
75 Ibid., p. 31.
76 Jay Miller, *Northrop B-2 Spirit* (Leicester: Midland Publishing, 1995), p. 1.
77 Miller, p. 2. Note that stealth technologies do not actually make an aircraft "invisible" to radar, but just significantly reduce the energy reflected back to the source radar. What this does is dramatically reduce the distance away from a radar site at which the aircraft can be located. Because a radar's detection ability decreases as the fourth-power of the distance, it is much harder to "hide" aircraft as they approach the radar site – a radar is 10,000 times as sensitive to targets five miles away as it is to targets fifty miles away. Therefore, even a plane like the B-2 cannot fly with total impunity, but it can come far closer to radar sites than non-stealth aircraft.
78 Right angles can act as corner reflectors that mirror a beam directly back toward its source; especially problematic are two adjacent right angles, like a bolt that sticks out slightly from an aircraft skin.
79 Miller, *Northrop B-2 Spirit*, p. 3.
80 Ibid., pp. 4–5.
81 Ibid., p. 235.
82 Ibid., p. 11. It was also discovered that various absorption materials can be layered to produce protection across frequencies.
83 R. G. P. Collison, "Fly-by-Wire Flight Controls," *Computing & Control Engineering Journal*, Vol. 10, No. 4 (August 1999), pp. 141–142.
84 Miller, *Northrop B-2 Spirit*, p. 5.
85 Michael A. Dornheim, "Fly-by-Wire Key to 'Pure Stealth'," *Aviation Week & Space Technology*, Vol. 132, No. 15 (April 9, 1990), p. 36.
86 Michael E. Brown, *Flying Blind: The Politics of the U.S. Strategic Bomber Program* (Ithaca, NY: Cornell University Press, 1992), p. 273.
87 Doug Richardson, *Stealth Warplanes* (New York, NY: Salamander, 1989), pp. 114–115.
88 Brown, *Flying Blind*, p. 295.
89 Richardson, *Stealth Warplanes*, pp. 150–151.

90 David A Fulghum and Douglas Barrie, "UCAVs Need to be Cheap and Survivable, So Costly Low-Observable Formulas get a Second Look," *Aviation Week and Space Technology*, Vol. 162, No. 14 (April 4, 2005).
91 Benjamin S. Lambeth, "Kosovo and the Continuing SEAD Challenge," *Aerospace and Power Journal*, Vol. 16, No. 2 (Summer 2002), pp. 17–18.
92 The creation of cooperative engagement capability, which allows a ship to use transmitters located elsewhere to engage a target helps overcome this problem. E-2 aircraft, which can look down onto incoming platforms and relay their location to ships, also help.
93 As of fall 2008, only two to three are planned. Megan Scully, "Negotiators Agree to Buy More F-22s, Zumwalt Destroyers," *Congress Daily*, September 24, 2008.
94 Christopher Cavas, "Is New US Destroyer Unstable?" *Defense News*, April 2, 2007.
95 Ibid.
96 On offense–defense theory, see Stephen Van Evera, *Causes of War: Power and the Roots of Conflict* (Ithaca, NY: Cornell University Press, 1999). For a critique, see Richard Betts, "Must War Find a Way: A Review Essay," *International Security*, Vol. 24, No. 2 (October 1999) pp. 166–198.
97 Defense Science Board, *21st Century Strategic Technology Vectors, Vol. II: Critical and Enabling Technologies* (Washington, DC: Office of the Secretary of Defense, 2007).

3 From conservatism to revolutionary intoxication

The US Army and the second interwar period

Colin Jackson

At end of the Cold War, the US Army confronted three challenges. First, the collapse of the Soviet Union ushered in an era of strategic uncertainty. The NATO Central Front scenario that had formed the anchor of the Army's doctrine, equipment, and force structure after Vietnam had disappeared. More than the other services, the Army faced existential questions about its future roles and missions. Second, the Army faced a number of external pressures: the technological opportunities and threats of the Revolution in Military Affairs (RMA), wartime experience, and bureaucratic competition for roles and funding. Third, the Army's modernization choices did not occur in a vacuum; they were guided by internal preferences regarding the proper role of the organization in the post-Cold War environment. The tension between the missions demanded by political authorities and the missions the Army preferred to undertake shaped the central decisions on interwar modernization.

Ultimately, internal preferences and the competition for relevance proved the most powerful drivers of the Army's behavior in the interwar period. The Army sought to preserve its own ideas about its place in the defense establishment. Between 1989 and 1998, this preservation effort brought a measured and incremental exploration of the RMA as it might contribute to improvements on the existing core model of the late Cold War army. From 1999 through 2007, the Army moved aggressively in the direction of an alternative vision of warfighting. This abrupt shift, brought on by threats of functional obsolescence, nevertheless bore the imprint of the Army's underlying preference for conventional warfare. Wartime experience and estimates of the likely demands on the Army played a far smaller role. Neither the consistent demand for low-intensity contingencies nor the experience of protracted counterinsurgency diminished the Army's attachment to a vision of a future dominated by "firepower-intensive, limited war."[1] In short, the Army's behavior in the second interwar period represented the triumph of institutional preference over experience and rational expectations.

The fate of Army modernization in the second decade of the twenty-first century remains uncertain. The leading Army modernization programs, the technologically ambitious Future Combat Systems (FCS) and the Modularity initiative, continue in spite of growing doubts about their rationales and affordability. The decision in 2006 to increase the size of the Army from ten to twelve division equivalents only amplified the pressure on these modernization initiatives. In the wake of Iraq, the Army will face at least three basic tradeoffs: capital investment versus labor expansion, the close fight versus rapid deployment and the deep fight, and conventional warfighting versus irregular warfare. The end of supplemental budgets will force the Army to choose between its commitment to ambitious capital and organizational modernization and the manpower expansion necessary to meet the future demands of low-intensity conflict.

The Cold War context: the post-Vietnam renaissance

With the end of major American involvement in Vietnam in the early 1970s, the Army shifted its focus from peripheral, low-intensity warfare to the problem of high-intensity combat in Europe. The reforms begun under Army Chief of Staff Creighton Abrams transformed the Army in doctrinal, structural, and technological terms. By the mid-1980s, the US Army had begun to field an entirely different force built around the AirLand Battle doctrine and a series of new combat platforms. This European army formed the nucleus of the force that would fight and win in Operation Desert Storm.

The post-Vietnam renaissance had four pillars. The first of these was the strategic problem itself – the conventional defense of Western Europe against Soviet ground invasion. Though this task was enormously challenging, it was neither novel nor unsettling. Transformation took place within the bounds of a familiar theater and represented a return to what the Army saw as its preferred mission – high-intensity, conventional warfare. The central front was also stationary – the Army could plan its defense and pre-position its equipment in anticipation of the struggle. For this reason, the problem of strategic lift and rapid deployment could be mitigated by careful prewar preparation.

The second pillar was a series of weapons systems programs known collectively as "the Big Five." General Abrams and his deputy General DePuy made a decision in 1972 to focus Army procurement funds on five major weapons systems: a main battle tank, an armored personnel carrier, a utility helicopter, an attack helicopter, and an air defense missile system.[2] These five programs, which would produce the M1 Abrams tank, the M2/3 Bradley Fighting Vehicle, the UH-60 Blackhawk and AH-64 Apache, and the Patriot Air Defense system, were the foundation of a complete recapitalization of the service.[3] Programs launched in the 1970s and funded on a massive scale from 1979 through the end of the Cold War gave the Army the physical tools to fight high-intensity warfare against a numerically superior Warsaw Pact.

The third pillar was the doctrinal revolution. In the wake of the 1973 Arab–Israeli War, General DePuy produced a new operations manual that laid out a new and deeply controversial concept of Army operations. Though many of the ideas contained in the 1973 draft were overturned in a later version of the manual, the DePuy initiative set in motion a series of doctrinal refinements that culminated in the 1982 and 1986 versions of the Army's new AirLand Battle doctrine.[4] Under the new concept, the Army planned to use ground forces in conjunction with air forces to defeat the oncoming Soviet forces in depth. The new doctrine put heavy emphasis on operational agility and overturned several decades of reliance on fairly inflexible, attrition-based forward defense in Germany.

The fourth pillar was the Combat Training Center (CTC) system. The Army used a combination of laser simulation and fully instrumented ranges to stage battalion-level, force-on-force engagements under realistic conditions. Whether in the desert terrain of the National Training Center (NTC) at Ft. Irwin, California, or in the hilly and forested terrain of the Combat Maneuver Training Center (CMTC) at Hoenfels, Germany, American units could sharpen their skills in the face of a skilled adversary trained and equipped to simulate the tactics of the Soviet army. The combination of realistic training and aggressive after-action reviews increased the tactical proficiency of American units at the battalion level.[5]

The circumstances of the Central Front and the Cold War more generally contributed to the success of Army modernization during this period. The task, though challenging, was simple and stable. The Army could afford to optimize its doctrine, equipment, and force structure around a single scenario. Equally important, that scenario was consistent with the Army's preference for conventional, high-intensity warfighting. The focus on Europe allowed the Army to build a force perfectly suited to a mission they accepted as uniquely important. Neither of these conditions held after the Cold War. In the post-Cold War era, the Army was forced to accept increased uncertainty and a growing gap between the constabulary demands of civilian policymakers and the service's preference for conventional war.

Dual shock: the end of the Cold War and Desert Storm

In the space of two years, the Army weathered two major shocks. The collapse of the Soviet Union in 1989 effectively eliminated the service's core mission. More than the other services, the Army's doctrine and structure had been designed around this single European scenario, and its sudden disappearance left the Army adrift amid mounting budgetary pressures. Less than nine months after the fall of the Berlin Wall, Saddam Hussein's invasion of Kuwait tabled the problem of interwar adaptation and forced the Army to apply its late-Cold War tools to a different challenge – conventional, limited war in the Middle East. Though Operation Desert Storm validated many of the technological and doctrinal elements of the Cold War model, the outcome of the war raised questions inside the Army about its intrinsic limitations. The Army's thinking on modernization in the interwar period was shaped by the perceived need to overcome these shortcomings, particularly in the areas of strategic and theater mobility.

The rapid collapse of the Soviet Union raised fundamental questions about the service's role and its likely opponents. While the Army began to explore these questions, the pressure for budget savings began to mount. Though the level of cuts remained unclear in 1990, the Army had to confront the possibility of deep reductions in end strength and the dilemma of strategic reorientation.

Even before Desert Storm, it was clear that the Army's center of gravity would shift from Europe to the continental United States, with a likely cut of several divisions.[6]

Operation Desert Storm, on the other hand, appeared to validate the training, doctrine, and equipment of the late Cold War. American ground forces smashed the Iraqi defenses and drove their army out of Kuwait in 100 hours. On the tactical and operational levels, American equipment and training provided a decisive overmatch in all major engagements. Though some Iraqi forces put up stiff resistance, American forces were able to inflict enormous damage at negligible cost. American casualties in the ground war totaled 140, roughly 20 percent of which were the result of fratricide rather than enemy fire.[7] The major disappointment of the campaign was the failure to encircle and destroy the Iraqi Republican Guard divisions as they moved north out of Kuwait into southern Iraq. Had Army forces managed to cut off the path of retreat, the coalition might have defeated these elite units and precipitated the collapse of the Iraqi regime. In the immediate aftermath of the war, however, the scale of the achievement dwarfed any perceived shortcomings. The Army was proud of its accomplishments and aware that the victory was the culmination of a two-decade restoration of the Army as a professional institution.[8]

Even in victory, some senior leaders worried about the Army's real and perceived performance. The speed of the Iraqi collapse meant that Army leaders had to combat the perception that the Air Force had defeated the Iraqis prior to start of ground operations. The Army's difficulties in the public relations arena were largely self-inflicted. While the Air Force had captured the popular imagination with images of precision bombing, the Army had restricted journalists' access and made the ground war nearly invisible to the national audience.[9] On both the operational and strategic levels, the Army's heavy forces were less agile than senior commanders would have liked. The inevitable delays in moving heavy mechanized forces from Europe and the United States to the Persian Gulf meant that for the first five weeks of Operation Desert Shield, the dismounted infantry of the 82nd Airborne Division were the only forces on the ground to block a possible Iraqi drive south into Saudi Arabia. Even after two corps' worth of heavy divisions had been assembled in Saudi Arabia, the logistical burden of sustaining the rapid advance limited the utility of the force on the operational level. Some combination of unwieldy command arrangements and logistical inflexibility appeared to have contributed to the escape of Iraqi Republican Guard units north into Iraq.[10]

The first decade: incremental change, constabulary entanglements and intellectual ferment (1989–1998)

In the first decade of the interwar period, the Army's response to transformation was profoundly conservative. Faced with deep reductions in force structure and considerable strategic uncertainty, the Army focused on the defense of readiness and the incremental exploitation of emerging technology. Though Army Chief of

Staff General Sullivan launched the Force XXI program to explore the use of information technologies, the initiative was focused on improving the capabilities of the existing divisional forms of the late Cold War period.[11] Later in the decade, Sullivan's successor, General Reimer, commissioned a more open-ended and ambitious study of future war in the Army After Next (AAN) program. Although the Army's operational focus was on Balkan peacekeeping missions, the AAN project began to sketch out a more ambitious vision of a high-technology army capable of rapid deployment and decisive action in "firepower-intensive, limited wars."

The Army's measured approach to transformation reflected the confidence in the existing force and concerns about the impact of force reductions. Shortly after assuming his post as Army Chief of Staff in 1991, General Gordon Sullivan identified a number of challenges facing the Army in the post-Cold War environment. The first, and clearly the dominant one in Sullivan's mind, was to "maintain the edge" the Army had honed in the Cold War: "... we must ensure a trained and ready Army at each point in time throughout the period of change. This means we must retain the warfighting advantage we demonstrated in Operations JUST CAUSE and DESERT STORM."[12] Sullivan was acutely aware that the Army had seldom succeeded in maintaining readiness during postwar reductions in force structure. In the four years following Desert Storm, the Army cut its active force structure from the Cold War peak of eighteen divisions to ten. Sullivan's mantra throughout his tenure as Chief of Staff was "No more Task Force Smiths," a reference to the collapse of the first army task force deployed to Korea in 1950.[13] For Sullivan, Task Force Smith was a parable: it demonstrated the costs of past failures to maintain training and material readiness in an environment of steeply declining budgets, plummeting end strength, and preoccupation with constabulary functions.[14] The Task Force Smith debacle also underlined the risks of a victor's complacency; the collapse in Korea at the hands of a third-rate power came fewer than five years after the stunning triumphs over Japan and Germany in World War II.

Though Sullivan's primary preoccupation throughout his tenure was the defense of readiness rather than the pursuit of technological transformation, he did sponsor limited experimentation with RMA technologies. His primary contribution to transformation was the Force XXI program. In March 1994, Sullivan directed the Army to conduct large-scale exercises with a "digitized" division at Fort Hood. Force XXI was an attempt to incorporate new digital technologies into an existing heavy division structure to improve command and control, and particularly "blue force" situational awareness. Of the three major RMA technologies, information technology was the core of the Force XXI initiative, while the role of stealth and precision technologies remained largely unexplored.

Though it took nearly three years to implement and test the digitized 4th Infantry Division at Fort Hood, the Army considered the Force XXI project a success. Though beset with countless teething problems, the integration of a tactical internet enabled the division staff to establish a much clearer idea of the location of its own units. Subsequent maneuvers indicated that increases in situational awareness of friendly forces allowed commanders to spread their units over a much larger area. These advances in command and control made tactical speed the new limiting factor in operational maneuver. On numerous occasions, practical limits on closing speeds prevented 4th Infantry Division commanders from exploiting promising opportunities revealed by intelligence assets.[15]

General Dennis Reimer, Sullivan's successor as Army Chief of Staff, sought to build on Sullivan's Force XXI experiments by launching a longer-range study "The Army After Next" (AAN).[16] Whereas Force XXI had been an explicitly evolutionary attempt to graft emerging technologies onto the existing force, AAN, led by Major General Robert Scales, was charged with a more revolutionary examination of the role of the Army in wars in the 2020–2030 timeframe. The project, which ran from 1995 to 1997, used a combination of strategic and operational war gaming to derive insights and develop operational concepts for future war. In Scales's view, the group had to lay the groundwork for the force and doctrine that would follow the inevitable retirement of Cold War-era equipment and leaders in the post-2010 environment.[17]

Scales's group emerged from four years of wargaming with a vision of a dramatically different force optimized for what Scales would later call "firepower-intensive, limited wars."[18] As Scales made clear at the time, the Army's modernization would be driven by a need to maintain the service's relevance in a changed interservice and threat environment: "If the Army is to remain relevant to the security needs of the nation we must begin now to accelerate the speed with which we can project legitimate, powerful, and balanced forces to threatened regions overseas."[19] Scales was deeply concerned with the time it had taken to deploy major ground forces to Saudi Arabia. In the month between the deployment of the lightly equipped 82nd Airborne Division and the arrival of the first heavy units of the 24th Infantry Division, the US military position had been terribly exposed. Had Saddam Hussein chosen to push south during that first month, the Army would have been entirely dependent on air support to stop the Iraqi forces. Ideally, the future army would eliminate this gap by combining rapid deployment with far more capable combat systems and doctrine. Desert Storm had sharpened the perceived contrast between an Air Force that was capable of rapid exercise of decisive force at great range and a lethal but ponderous ground force that took six months to assemble and deploy. Seen in this light, AAN was an argument for the continuing relevance of the Army in future expeditionary warfighting. In order to preserve its place, and avoid simple relegation to constabulary missions and peacekeeping, the Army had to develop a way to project force rapidly and produce decisive results.

The core of the AAN vision was a force restructured around light, high-technology, combined arms units capable of rapid deployment from bases in the continental United States. Smaller (brigade- rather than division-sized), "balanced" units would deliver precision fires and provide dominant maneuver to collapse the will of their opponents. Units would be air transportable by design, enabling them to move into theater at great speed and then bound within theater using tilt rotor aircraft and C-130s.[20] This notion of "air mechanization"[21] was seen as the next logical step in expanding the mobility gap between the Army and its likely opponents. Advances in active protection systems (APS) and weapons technology and would enable these mobile light units to be both survivable and lethal. The dominant situational awareness provided by advanced sensors and networking would substitute for the past reliance on heavy armor and enable units to engage and destroy the enemy at a distance. Writing in 1998, Scales explains the transformation in ground combat in these terms:

> All advanced ground vehicles would rarely be required to face main battle tanks head on, which makes it possible to limit their weight by reducing the need for heavy armor. They will survive through a combination of speed, agility, active protection, signature management and control, comprehensive situational understanding, terrain masking, deception, and indirect fire. Greater ground speed on and off roads will be possible because of advanced suspension systems, power trains, and engines. Greater fuel economies will result from significant weight reduction and advanced propulsion system designs.[22]

Anticipated rapid changes in technology would set up a virtuous circle in which increasing situational awareness drove platform weight down and operational and strategic mobility up. What Scales and his contemporaries were describing was a fundamental shift in the nature of ground combat brought on by the application of the full suite of RMA technologies: stealth,[23] precision, and information technology. For centuries armies had relied on the "movement to contact" to locate and engage their opponents. Scales argues that advanced C4ISR would make this type of meeting engagement far less frequent. In his words, "the close fight was becoming less close,"[24] and this future force might be able to "collapse the will" of the opponent without the costly close battle of the past.[25] The disappearance of close battle made a move to light vehicles possible, and this in turn made air mechanization and a step change in operational mobility feasible. The AAN project was above all a capital and technology-intensive vision of future war, in which expeditionary force projection, high mobility, and deep attack would be the keys to victory. For Scales, the future path of Army modernization was clear: "America will remain dependent on its ability to continue to develop the technical means for substituting killing power for manpower."[26]

In retrospect, one of the most intriguing aspects of the AAN initiative was its treatment of low-intensity conflict. The AAN force was optimized for short, decisive campaigns against conventional opponents in distant theaters. In part this was an attempt to restore the Army's relevance as a national instrument of warfighting. It also appears to have been an attempt to redefine the future army in ways that arrested or reversed the recent slide toward constabulary missions. In his 2003 book *Yellow Smoke*, General Scales draws a clear distinction between the "firepower-intensive, limited wars" and "engagements below the threshold of physical violence:" humanitarian assistance, peacekeeping, peace enforcement, and constabulary missions.[27] Scales goes on to argue that an excessive focus on these missions would undermine conventional warfighting capabilities. The answer, he argues, is aggressive, conventional modernization:

> Decades of practical experience demonstrate convincingly that the US Army is good at peacekeeping because troops properly trained to fight full-scale war always perform well in less demanding contingencies. However, experience with the performance of troops thrown into combat in Korea, after years of constabulary service in Japan, suggests that the reverse does not apply. Very painful experience in the Korean War also amply reinforces the truth, that, in this new age of limited warfare, the conduct of shooting wars is what the United States most often does least well.[28]

Scales's argument is intriguing in two respects. First, the categorical distinction between war and "engagements below the threshold of physical violence" omits the category of manpower-intensive and firepower-limited, internal war (counterinsurgency) that would dominate the operational landscape from 2003 through 2007. Scales's categorization, which was representative of contemporary thinking among Army futurists, strongly suggests that the same theory of victory could be applied to all shooting wars. Second, Scales clearly asserts that low-intensity conflict is the lesser included case; a force trained for high-intensity conflict could handle low-intensity conflict, but that a constabulary army could not perform in a high-intensity environment.[29] The real risk in Scales's opinion was a neglect of high-intensity capabilities. The overwhelming emphasis on high-intensity war and the tendency to minimize the challenge of low-intensity conflict would have a significant impact in the second interwar decade as the Army attempted to grapple with four shooting wars that bore scant resemblance to those envisioned under the AAN.

Although Scales and the AAN planners sought to articulate a vision of the Army as an agile, capital-intensive warfighting force, civilian policymakers were busy employing the Army in a series of less glamorous, largely static, and labor-intensive constabulary missions. The conclusion of the Dayton Peace Accords in the winter of 1995 paved the way for the deployment of 20,000 US Army troops into Bosnia. The initial deployment from Germany into Bosnia included an embarrassing three-week delay in bridging the Sava river, an episode that highlighted the mobility problems with the heavy force.[30] Once in Bosnia, the American forces saw their readiness in core warfighting missions erode; troops kept busy with presence patrols and stabilization duties could not maintain their proficiency in combat skills. The peacekeeping missions were not popular with officer corps or the rank and file, with many troops referring to Military Operations Other Than War (MOOTW) as "Military Operations Other Than What I Signed Up For." More troubling from the point of view of the Army was the essentially open-ended nature of these missions; given a one-year rotation policy, the maintenance of a division in Bosnia absorbed most of the available manpower in the European theater. The Bosnia mission came to stand for everything the military would prefer to avoid: open-ended, constabulary duty that drained resources, eroded readiness, generated bad publicity, and depressed morale.

The contrast between the AAN vision of short, decisive limited wars and the contemporary drudgery of the Balkans captured the basic tension of the first decade: the clash between policymakers' demands and the Army's preferred missions. Army leaders recognized the need to satisfy policymakers' demands for flexible response to contingencies, but they bridled at the notion that responsiveness might lead to permanent constabulary duty. Those who defended the relevance of ground power to American policy generally did in it ways that sought to move the Army away from MOOTW and toward the more desirable category of high-intensity, limited war.

In the first decade of the interwar period, the Army took a conservative approach to modernization. Army leaders tried to maintain readiness in the face of major budget cuts and force reductions. The Force XXI project was emblematic of the Army's evolutionary approach to the RMA. While the Army did engage in more radical and open-ended speculation about the future application of RMA technologies to ground combat in the AAN project, these ideas were largely confined to concept development and experimentation.

The second decade: revolutionary intoxication – Kosovo, future combat systems, modularity, and Iraq (1999–2007)

If the first decade of the interwar period was dominated by a conservative approach to modernization, the second decade would prove quite the opposite. Stung by its inability to contribute to the success of the war in Kosovo, the Army announced an ambitious plan to revolutionize its forces and doctrine. Impelled by bureaucratic pressures more than direct experience, the Army set out to accelerate the implementation of the AAN vision. Although the service acknowledged the need to prepare for the full spectrum of contingencies, the three major service initiatives that emerged from this impulse – the Stryker Interim Brigades, the FCS, and modularity – were clearly optimized for "firepower-intensive, limited war." Though this emphasis on conventional war was clear, the Army adopted Scales's argument that a force designed for high-end conflict would be fully capable of tackling low-intensity missions.

Just as the Army embarked on this sweeping modernization program, four wars tested the Army's existing force and its vision of future war. A service that had embraced transformation as a way of resurrecting its relevance soon found its services in high demand. The problem was that these new wars fit neither the existing template of constabulary occupation nor the imagined framework of firepower-intensive, limited war. Though major combat operations in Afghanistan and Iraq highlighted the significance of rapid deployment and precision technologies, the later phases of both conflicts demonstrated the limits of high technology in ground operations and the continuing need for manpower and heavy protection in close fighting.

By 2007, the Army's modernization programs and the needs of the counterinsurgent (COIN) campaigns in Iraq and Afghanistan had begun to pull in opposite directions. In spite of serious empirical challenges to core assumptions of the FCS and modularity initiatives, the institutional Army clung to a program of capital-intensive modernization for future conventional war. Experience in Iraq and Afghanistan suggested that the problem was not capital but labor. Though supplemental budgets had enabled the Army to continue to pursue FCS and modularity, the Army Chief of Staff's reluctant acceptance of a major increase in force levels in 2007 represented an important break with the earlier commitment to capital-intensive modernization. Recent wartime experience also undermined the notion that improved situational awareness could replace reliance on heavy armor. Even in the relatively primitive threat environment of Iraq, American units found that passive armor protection was the only reliable response to the leading threats off-route mines and infantry anti-armor weapons.

Kosovo, Task Force Hawk, and the threat of functional obsolescence

More than any other single event, it was the Kosovo war that convinced the Army of the need to pursue radical modernization. Although Kosovo had been a "firepower-intensive, limited war," the Army had proven unable or unwilling to contribute to the outcome. The war became a showcase for the independent power of air forces and the risk aversion and inflexibility of the Army. It was the bureaucratic aftershocks of the Kosovo war that compelled the Army to move from evolutionary change of Force XXI to the immediate implementation of a more radical AAN vision.

The structure of the Kosovo war favored the Air Force. The Clinton administration's limited objectives and casualty aversion made "air only" strategies more appealing. NATO's broad goals in the Kosovo campaign were to force an end to the campaign of ethnic cleansing and compel the Serbian government to accept Kosovar autonomy.[31] NATO embarked on what was to be an eleven-week air campaign to compel Serbia to accept these demands.

Even before the start of military operations, however, the NATO Commander, General Wesley Clark, had developed a range of ground options to guard against the potential failure of the air campaign. The first option involved the deployment of a task force of US Army AH-64 Apache attack helicopters and long range ATACMS missile batteries to Macedonia.[32] Clark argued that the Apaches would be an important complement to USAF fixed-wing air strikes, giving the NATO command the ability to take out small and mobile targets that might escape the high-altitude, precision-guided munitions (PGM) attacks.[33] The Apaches would cross into Kosovo under cover of night and after artillery and fixed-wing air assets had suppressed Serbian air defense assets.[34] The ATACMS missiles, with their long range (~200 miles) and short time of flight (two minutes), would suppress enemy air defenses and give NATO a means to attack fleeting targets not only in Kosovo but in Serbia proper. Clark was emphatic that the Apache/ATACMS option, which grew into Task Force Hawk, would not involve any ground penetration of Kosovo. Instead, Clark wanted to use these strike assets to complement the air strikes by fixed-wing aircraft and cruise missiles.[35] Since all of the units involved were already stationed in Germany and could be air deployed, Clark anticipated that it would be possible to push the 1,800 troops, the aircraft, and the equipment into Macedonia in less than two weeks.[36]

General Clark also began to put together a more ambitious ground invasion plan. Under this plan, NATO ground forces would move from bases in Macedonia and Albania into Kosovo to defeat and eject the Serbian ground forces from Kosovo.[37] Though the NATO staff had started with a planning figure of 70,000 NATO troops, subsequent revisions to the Serbian order of battle led Clark to insist on a total force of between 200,000 and 250,000 troops.[38] Under the revised plan, the US would contribute 175,000 personnel, including some 100,000 combat troops.[39] Clark estimated that it would take seventy-five days to move the six US divisions and support troops into position to launch the ground invasion.[40]

Clark's plans, both for Task Force Hawk and the ground invasion plan, ran into immediate resistance from the Joint Chiefs of Staff and particularly the US Army Chief of Staff, General Reimer. The Army staff raised two major objections to the Task Force Hawk deployment. First, Reimer and others feared that Task Force Hawk's cross-border raids might provoke Serbian retaliation in the form of artillery and rocket fire into Macedonia or Albania. Second, Army leaders considered the employment of the Apaches inordinately risky. General Reimer and others feared that the Apaches would be decimated by Serbian air defenses and that the proposed target set did not justify the additional risk.[41] According to Clark, Army sources communicated these doubts to the White House directly, apparently suggesting that 50 percent of the Apache force might be lost to enemy fire.[42] While General Clark repeatedly pressured senior Army leaders and the CJCS General Shelton to endorse the use of the Apaches, the ongoing Army resistance fatally undermined military and White House support. On May 18, 1999, with the air campaign underway, President Clinton publicly ruled out the use of the Apaches then in place on the grounds of excessive risk.[43]

Though the Army resisted Clark's ground options, the practical difficulty of the Task Force Hawk deployment far exceeded expectations. Concerns about Serbian retaliation led Army planners to increase the number of covering forces attached to Task Force Hawk and shift the base of operations from Macedonia to Albania. What had started as a proposal to send a 1,800-man helicopter/artillery task force to Macedonia soon ballooned into a commitment to deploy a 5,500 man task force, including three tank and mechanized infantry companies, to Albania by air.[44] Clark's plans for the ground invasion option met with similar resistance and delays.[45] Those discussions never reached a final conclusion, with the Army's reluctance to commit ground forces to either venture the defining feature in both cases.

The objections of military leaders and the practical obstacles to deploying heavy forces into unimproved military facilities in Albania and Macedonia impeded the deployment of Task Force Hawk. Whereas Clark had hoped to have the Apaches in place and ready to operate within seven to ten days, the actual deployment of Task Force Hawk, from formal orders to full operational capability, took forty-four days.[46] Once in place, the forces had to overcome the limitations of the Albanian air base at Triana-Rinas. Delays in reaching fully operational capability and the loss of two Apache helicopters in separate training accidents reinforced concerns among senior military and civilian leaders in Washington. On May 26, 1999, three weeks after the task force had reached fully operational capability, the Joint Chiefs of Staff and the Secretary of Defense formally rejected General Clark's request to use the Apaches in support the campaign.[47]

The Kosovo campaign ended on June 9, 1999, with the Serbian acceptance of NATO's major demands. This success, achieved without the use of TF Hawk or the commitment of major ground forces, cast increased doubt on the utility of the Army in the post-Cold War era.[48] The specific merits of the arguments over Apache use and deployment problems were almost irrelevant.[49] What was clear by the end of the Kosovo campaign was that the Army was the service that said no. The Deputy Secretary of Defense, John Hamre, expressed this frustration with the Army in a September 1999 public interview with *Defense News*: "If the Army holds onto nostalgic versions of its grand past, it is going to atrophy and die."[50]

This public slap in the face, coming just as General Eric Shinseki prepared to replace General Reimer as Army Chief of Staff, appears to have been the turning point in Army interwar transformation. Fearful that Kosovo had made the Army appear obsolete, Shinseki was prepared to embark on a far more sweeping set of changes. Only by making the Army more responsive to policymakers could Shinseki guard against further threats to the service's roles, missions, and resources.

General Shinseki and the embrace of transformation

Faced with mounting criticism of the Army's utility, Shinseki seized on the AAN as a template for accelerated transformation. Taking what had been an ambitious vision for war in 2020, Shinseki proposed a series of steps to create a force capable of rapid deployment and decisive action in distant theaters. Shinseki set an ambitious timeline for the fielding of an "Interim Force" of light, wheeled vehicles and for the accelerated development of an "Objective Force" capable of realizing the full vision of the AAN planners.

Though Shinseki's announcement predated the Bush administration and the shock of 9/11, it was the AAN vision of "firepower-intensive, limited war" that would set the modernization trajectory for the remainder of the decade.[51] Transformation, as embodied in three major initiatives – the Stryker Interim Force, the Future Combat Systems, and modularity – would survive the experience of an Iraq war that undermined many of its core assumptions. The Army's response would continue to be one based on capital-intensive, conventional war. The post-9/11 era would be dominated by the tension between heavy demand for the Army manpower in counterinsurgency and the service's attachment to a modernization program that provided high technology solutions to future conventional wars.

Four months after the end of the Kosovo campaign and only a month after the public rebuke by Secretary Hamre, the newly minted Army Chief of Staff

Shinseki announced his plans to accelerate modernization in a speech to the Association of the United States Army. Shinseki argued that the Army must pursue a much lighter, more technologically advanced, air-deployable force designed in order to provide policymakers with a genuine ground power option in the post-Cold War era:

> We will prioritize solutions which optimize smaller, lighter, more lethal, yet more reliable, fuel efficient and more survivable options. We will seek technological solutions to our current dilemmas. We want the best combination of technologies that will provide survivability through low-observable, ballistic protection, long-range acquisition, deep targeting, early attack and first round kill at smaller caliber solutions. Can we, in time, go to an all-wheel vehicle fleet where even the follow-on to today's armored vehicles can come in at 50 percent-70 percent less tonnage? I think the answer is yes, and we're going to ask the questions and then go where the answers are. With the right technological solutions, we intend to transform the Army, all components, into a standard design with internetted C4ISR packages that allow us to put a combat-capable brigade anywhere in the world in 96 hours once we have received execute liftoff, a division on the ground in 120 hours, and five divisions in 30 days. Being able to do so gives the National Command Authority a genuine deterrent capability – when ordered, we intend to get to trouble spots faster than our adversaries can complicate the crisis. Once there, we intend to leverage for de-escalation and a return to stability through our formidable presence. But if deterrence fails, we will be postured to prosecute war with an intensity that wins at least cost to us and our allies and sends clear messages for all future crises. And when technology permits, we will erase the distinctions, which exist today, between heavy and light forces and review our requirements for specialty units.[52]

Shinseki's speech was an unequivocal embrace of technological transformation. All major elements of the AAN vision were present: the objective of greatly increased strategic and operational mobility, the reliance on RMA technologies to reduce weight, and the move from the existing heavy and light division structures to a standard brigade design. More important, Shinseki explicitly tied these changes to the problem of policy relevance. As such, his remarks were a direct response to the disappointment of Task Force Hawk and a demonstration of the Army's willingness to change. The only major departure from the AAN framework was the greatly accelerated timeline and the inclusion of an "Interim Force" to bridge the gap between the current force and the "Objective Force" of the AAN.

In later elaborations of his vision, Shinseki sketched out a force divided into three categories: the legacy force, the interim force, and the objective force. The legacy force consisted of the existing major combat formations and equipment of the late Cold War. In Shinseki's view, these platforms and units lacked the strategic and operational mobility to respond promptly to future contingencies. The objective force would be reorganized around the Future Combat Systems suite of networked, light, and air-transportable platforms. The interim force was Shinseki's bridging concept. By fielding several brigades designed around an off-the-shelf wheeled vehicle, Shinseki hoped to build more capable rapid deployment forces and test the technical and organizational concepts proposed for the "Objective Force."[53] The fast-track approach to fielding the interim force would, like the FCS and modularity initiatives, reveal the dangers of rapid modernization.

The arrival of the Bush administration and Secretary of Defense Rumsfeld only intensified the pressure to transform. The new Secretary entered the Pentagon with a conviction that the Army was philosophically and functionally ill-suited to the problems of future war. Early face-offs between Secretary Rumsfeld and General Shinseki and Secretary of the Army Thomas White over the Crusader artillery system only increased the tension between the Secretary and the Army senior leadership. On the eve of 9/11, Rumsfeld's team went so far as to consider eliminating two Army divisions in an effort to redistribute resources and hasten the advent of the RMA.[54]

Stryker and the interim force

The first step in Shinseki's transformation plan was the fielding of a medium-weight, wheeled force of six brigades based on off-the-shelf technology.[55] Following a competition in 2000, the Army selected the General Dynamics Land Systems Piranha LAV-III eight-wheeled vehicle as the basis of its Stryker family of vehicles. Like the more ambitious FCS program, the Stryker program was to be driven by two imperatives: a need to limit platform weight to meet C-130 transportability requirements (18-ton maximum), and a commitment to improved situational awareness as a substitute for passive armor protection.

The rapid fielding of the Stryker brought inevitable teething problems. While the Stryker was based on the proven Mowag LAV design, the Stryker vehicle, even in its simplest IFV configuration, was essentially a prototype rather than a production system. Though the Army sought to minimize the risks of bypassing traditional testing and evaluation, the accelerated fielding process revealed a series of shortcomings with the computer systems, weapons stations, and other components.[56] While none of these design flaws proved fatal, the need to redesign major components in the midst of combat deployment revealed the costs of fielding an immature system.

More troubling were the problems that could be traced to fundamental design tradeoffs. The most basic of these was the mobility/survivability tradeoff. The requirement for strategic and intra-theater air mobility proved extremely challenging in several respects. First, the Stryker vehicle, even in its infantry carrier variant, exceeded the standard USAF volume restrictions for C-130 transport. Though the USAF waived its safety restrictions, and the Army managed to demonstrate the air transportability of a stripped down Infantry Fighting Vehicle variant under ideal conditions, the level of effort involved in making the Stryker air-transportable highlighted the limits of "air mechanization" with the interim force. Second, the weight of the Stryker, listed as 38,000 pounds, left little room for subsequent growth in platform weight or less-than-ideal flying conditions.[57] As a General Accounting Office study in 2004 revealed, even a 2,000-pound increase in platform weight would cut the practical C-130 range from 860 to 500 miles (roughly half of the Army's stated requirement of 1,000 mile range).[58]

The drive to reduce platform weight created a distinct survivability problem. Though the FCS program anticipated major strides in active protection systems (APS) to offset the loss of heavy armor, the Stryker vehicle offered only light armor protection against shell splinters and 14.5 mm fire.[59] Absent additional active or passive protection, the Stryker was vulnerable to light cannon fire, improvised explosive devices, and nearly all man-portable anti-armor weapons, including the ubiquitous RPG-7. The limits of a vehicle with this level of protection became clear in 2003 with the first deployment of a Stryker brigade to Iraq. In order to guard against RPG-7 fire, the Army had to install additional slat armor weighing 5,000 pounds. Even this limited solution[60] imposed major strains on the engine, drive train, and tires,[61] and significantly reduced the off-road mobility of the vehicle in wet conditions.[62] The addition of slat armor meant that the vehicle could not be transported by C-130 in a combat-ready configuration. The challenges in fielding even a modestly capable interim vehicle gave some indication of the hurdles ahead for the FCS program.

The insignificance of experience: Afghanistan, Iraq, and transformation

Just as the Army embarked on an ambitious plan to reconfigure the entire force for "firepower-intensive, limited war," the wars in Afghanistan and Iraq produced challenges that fell outside the scope of the Army vision. Though the opening phases of the campaigns in Afghanistan and Iraq highlighted the importance of certain RMA technologies, they did not necessarily support the specific vision articulated by the AAN. In Afghanistan, precision airstrikes and special operations forces appeared to eliminate the need for conventional mechanized forces. While the speed of the campaign underlined the importance of rapid deployment, the specific contribution of an FCS-type mounted force in a war with a Taliban-style opponent remained unclear. In the major combat phase of the Iraq war, improved C4ISR eliminated many of the fratricide problems seen in Desert Storm but did little to improve red force awareness or eliminate the need for a close combat capability. Irregular opponents in cluttered and densely populated urban environments limited the effectiveness of American sensors. As a result, in spite of considerable advances in sensor technology, American units tended to encounter the enemy in classic meeting engagements. In these engagements, it was the traditional combination of passive protection and close-in firepower that counted most.

The transition from major combat operations to counterinsurgency made the gap between Army modernization plans and the threat environment more pronounced. In Afghanistan and Iraq, US forces faced a challenge that was below the threshold of "firepower-intensive, limited war" but well above the "threshold of physical violence." In this middle category, entirely missing from the AAN vision, commanders needed a well-protected but manpower-heavy force. The threat of improvised explosive devices (IEDs) and rocket-propelled grenades (RPGs) led to the up-armoring of light platforms, including the HUMVEE and Stryker, in an effort to provide crews with a means to survive the opening shots of unpredictable, short-range meeting engagements. Legacy platforms such as the M1 Abrams tank and the M2 Bradley Fighting Vehicle offered US forces a means to deliver short range fires from a well-armored platform – a capability largely missing from the Stryker units and the planned FCS platforms. More significant still, the demand for large numbers of dismounted troops undermined the core assumption of the transformation project – the substitution of capital for labor. While General Scales had confidently asserted in 2003 that "America will remain dependent on its ability to continue to develop the technical means for substituting killing power for manpower,"[63] the success of Iraqi insurgents from 2003 to 2007 exposed an important vulnerability with no clear technological solution. As the Army struggled with counterinsurgency in Iraq and Afghanistan, it was hard to imagine that future opponents would elect to fight the Army on its preferred terms of "firepower-intensive, limited war."[64]

Full steam ahead: FCS, modularity, and Iraq

The most striking aspect of the Army modernization program between 2002 and 2008 has been its resilience in the face of disconfirming evidence from Afghanistan and Iraq. The modernization initiatives begun by General Shinseki in late 1999 in response to the Kosovo missteps have survived major empirical challenges to their underlying assumptions. Having guessed wrongly about the shape of war in the first decade of the twenty-first century, the Army has been remarkably resistant to amending or scrapping its preferred vision of future war. Buoyed by supplemental funding that absorbed the operational cost of the wars in Afghanistan and Iraq, the Army has continued to argue that the combination of an interim Stryker force, FCS, and modularity will be the appropriate response to the new threat environment. Only in 2007 did Congressional pressure to increase Army manpower raise the possibility of an imminent collision between the demands for capital modernization and end-strength expansion.

The FCS program was the core of Shinseki's vision for the Objective Force. The program was designed as a "system of systems" in which eighteen manned and unmanned light platforms (less than 20 tons) were networked together to provide high interoperability and dominant battlespace awareness.[65] Like the Stryker program, the FCS approach raised issues of affordability, technical risk, and mission focus. Unlike Stryker, the FCS program bundled a series of different, technologically challenging platforms and C4ISR systems together under a single super-program. None of the developments in the program or the contemporary threat environment resolved the basic uncertainties in these three areas. Instead, evidence appeared to suggest a need to restructure the program and perhaps to revisit fundamental operational assumptions and the technical and economic tradeoffs underpinning the program.[66]

Though the Army had billed FCS as its flagship program, major increases in the estimated cost of the program between 2003 and 2008 have raised questions of affordability. Starting from a planning figure of $91.4 billion in 2003, the estimated cost of the FCS program had ballooned to between $203 and 234 billion by 2007.[67] With an annual program cost of between $10 and $16 billion,[68] and a brigade equipment cost of roughly $6.7 billion,[69] the FCS program threatened to crowd out all other investments in technological modernization.[70]

The substantial technical risks inherent in the overall program and its subsystems made the proposed course yet more daunting. The vision of FCS demanded simultaneous technological solutions to the problems of air and ground mobility (strategic, theater, and mechanized) and survivability (situational awareness and APS as substitutes for armor). To meet a production date of 2012, the Army elected to push the FCS program into system development and demonstration (SDD) in 2003. At that time, just one of the thirty-one critical technologies associated with the system had been judged to be mature.[71] Although the Army argues that these risks can be managed, the program has tended to pass critical development gates with relatively immature critical technologies. As the Congressional Budget Office report of August 2006 noted, the ambitious technological goals raised the issue of systemic program risk:

> Such setbacks [in the maturity of major communications subsystems] reinforce the conclusion by GAO and the Army's independent review team that the schedule established for developing all eighteen FCS components in the time DOD typically takes to develop a single system is a risky one.[72]

Program success depended on the resolution of a vast number of critical and immature technologies. The failure of some number of these might deprive the force of the technologies needed to produce a deployable, lethal, and survivable force. The most likely effect of major technological setbacks would be to undermine the survivability of the force in likely threat scenarios.

External reports and other commentary by military officers and defense experts raised the more basic question of operational focus. In the wake of the Iraq experience, it was by no means clear that strategic and intra-theater air mobility was the leading problem facing the Army in its current and anticipated scenarios.[73] In the American political system, delays in political decision-making were generally more important than the technical or logistical limits on rapid force deployment. Even if strategic deployability were the leading problem, it remained unclear whether FCS could address it. As the CBO report highlighted, a transition from the current brigade equipment mix to an FCS brigade structure would reduce the deployment time by 19 percent at most.[74] An FCS brigade combat team would still require between 340 and 370 C-17 sorties to deploy, leaving brigade deployment times of between eighteen and twenty days rather than the ninety-six hours proposed by Shinseki. CBO analysts estimated that it would take 115 days to transport an FCS-equipped, four-brigade division to Djibouti by air.[75] Simply put, even FCS's dramatic reductions in platform weight did not translate into major improvements in strategic mobility. Because the cost of this marginal increase in mobility was a significant reduction in close-combat survivability, FCS, and air mechanization might reduce the capability of the future force.

The AAN concept was predicated on a shift toward longer engagement ranges and greater standoff. The Iraq war exposed two major problems with this assumption. First, experience in major combat and the counterinsurgency operations that followed indicated that improvements in C4ISR were unlikely to eliminate the close battle and the meeting engagement. The nature of irregular warfare in urban terrain meant that meeting engagements were still the most common way that Army forces entered battle. Second, the shift from the close fight to the deep battle would move the Army from its traditional mission space into one already well served by the USAF. Although the move from close combat to the deep fight might reduce the risk to friendly forces, it would be extremely contentious in interservice terms. At the same time, it was unclear how the light force would fufill the traditional close-combat mission should it arise. The advocates of FCS tended to deny the tradeoffs in the FCS equipment design, but the experience of close combat in Iraq tended to reinforce the ubiquity and difficulty of the close-combat problem – an area that no other service could adequately fill.

There is an alternative explanation for the Army's attachment to FCS. According to this line of reasoning, the Army never intended to build the futuristic force proposed under FCS. Instead, the Army planned to use the FCS program as a placeholder for future modernization. Stung by the cancellation of smaller programs (Crusader, Comanche), and impressed by the other services' success in guarding their leading procurement programs (F-22, F-35), the Army may have concluded that the best defense lay in large program size. If FCS could be made big enough, then the program could be used to shepherd modernization funds until the Army determined what it wanted to do on capital modernization. The only problem with this bureaucratic rationale was cost. The Army had spent $12 billion on FCS between 2003 and 2008 and Boeing, as the lead systems integrator, would extract significant fees regardless of whether the systems were successfully fielded.[76] Because any follow-on modernization program would involve additional research and development spending, the cost of the placeholder strategy was very high.

Modularity

In 2004, Army Chief of Staff Peter Schoomaker announced the introduction of a major reorganization of the Army force structure. The initiative, known as modularity, was the organizational analogue to technological transformation under FCS. Under modularity, the Army would reconfigure its forces around an increased number of standardized brigade combat teams, giving the national command authority a more effective and flexible force to meet the full range of contingencies. Though billed as a tool for "full-spectrum operations," modularity, like FCS, was a means of transforming the Army into a more lethal tool for conventional, limited war. The clash of imagined war and the contingencies of 2002–2007 raised much the same set of questions that beset FCS. Was modularity the right answer for "full-spectrum operations"? Was modularity affordable in light of the rising costs of FCS and demands for increased force levels?

Modularity was first and foremost about replacing divisions with brigades. According to proponents, the reorganization of the ten divisions into forty-two to forty-three brigades would improve unit effectiveness in several ways. First, a more robust brigade structure would provide national theater commanders with the appropriate unit size for contingency operations.[77] According to proponents, the experience of the 1990s had shown that divisions were too large and unwieldy for rapid deployment, and that brigades, armed with advanced weapons and reinforced with additional staff and intelligence assets, were the appropriate "unit of action" in the post-Cold War world. Second, modularity was about squeezing a larger number of deployable combat brigades out of a given force structure. According to the Army's calculations, the transition to a modular structure would increase the combat power of the Army by 30 percent and the rotational pool of deployable units by 50 percent. Because the pool of deployable units would grow, the number of times a given unit would have to deploy would drop; active units would spend two years in garrison for every year deployed, while Reserve and National Guard units would spend four and five years in garrison for every year deployed.[78]

Modularity also promised to address the problems of readiness and morale in an environment of high deployment demands. Stable, combined arms brigades would have an opportunity to train together, improving readiness and cohesion in advance of deployment. By stabilizing personnel in their brigades for longer periods of time, the Army leadership also hoped to reduce the strains on families and increase morale and retention.

Though modularity lacked the technical risks of FCS, it shared the problems of mission fit and cost. Like FCS, modularity seemed designed to maximize the effectiveness of the Army in firepower-intensive, limited wars. A 2006 study by the Institute for Defense Analysis (IDA) pointed out that the new brigade structure actually decreased the number of available maneuver battalions in each brigade by one-third, substituting a Reconnaissance, Surveillance, and Target Acquisition (RSTA) squadron for the traditional third maneuver battalion.[79] While the Army argued that the armored RSTA squadron was in fact more capable than the maneuver battalion it replaced, and provided additional advantages in the form of increased situational awareness, the disagreement between IDA and the Army highlighted the tension between those who saw maneuver battalions as the measure of capability and those who saw greater value in a reconnaissance-heavy, high-technology structure. Although the Army's claims on the value of RSTA battalions were defensible within the context of the imagined war scenarios, they were far less convincing in the context of lower intensity conflict, including counterinsurgency. In those conflicts, the coin of the realm was the maneuver unit or the foot soldier, and the modular redesign appeared to reduce the number of personnel available in a given brigade to accomplish those missions.

The cost of modularity was another leading concern. Though the Army had put forward a cost estimate of $28 billion in 2004 to cover an expansion from thirty-three combat brigades to forty-eight between fiscal year (FY) 2004 and FY 2011, their own estimate had grown to $48 billion by 2005. According to GAO estimates, even a move from thirty-three brigades to forty-two or forty-three would cost $75.5 billion over five years including personnel costs, construction, and equipment.[80] If the Army eventually fielded forty-eight brigade combat teams, then the total cost of modularity might approach $86.2 billion.

The collision of wants and needs: FCS, modularity, recapitalization, end strength, and Iraq

Through most of the second decade, the Army sought to defend its two primary modernization initiatives at all costs. In spite of the increasing demands of the wars in Iraq and Afghanistan, General Schoomaker remained adamantly committed to modernization, even if this meant a reduction in the size of the force.[81] Capital modernization and increasing capital intensity remained the central theme of modernization in the second decade of the interwar period.

By 2007, however, the Army's desire to maintain operations in Iraq and pursue two very costly modernization programs had begun to collide. First, in late 2006, the Army bowed to pressure and agreed to a major increase in end strength – a complete reversal of its earlier claims that capital modernization would take precedence over expansion in force levels. The proposed increase of 65,000 troops would move the Army from its initial projection of forty-two to forty-three BCTs to forty-eight BCTs by FY 2013. This increase would be costly; the Army projected the cost at some $70 billion over five years, of which $18 billion would be devoted to new equipment.[82] Second, the Army continued to request large supplemental funds to refurbish or replace equipment consumed in Iraq. The dramatically higher tempo of combat operations had begun to eat into the useful lives of a number of legacy platforms, from helicopters to tanks and armored personnel carriers. The cost of remanufacturing these platforms was forming a bow wave largely unincorporated in long-term Army budgeting. These "reset" costs rose dramatically over the course of the Iraq war: $1.3 billion (FY 2003), $3.7 billion (FY 2004), $6.5 billion FY 2005), $9 billion (FY 2006), and $17.6 billion (FY 2007).[83] Third, commanders in Iraq and Afghanistan continued to request additional equipment necessary to continue counterinsurgency operations.[84] Mine-resistant vehicles, up-armored HUMVEEs, body armor, weapons, and the like contributed to unfunded requests for FY 2008 alone of some $10.2 billion.[85] Defense Secretary Gates' decision in May 2007 to make the accelerated acquisition of Mine-Resistant Ambush Protected (MRAP) vehicles the Pentagon's number one acquisition priority added further pressure to the Army acquisition budget. The November 2006 plan to purchase a total of 4,100 MRAP vehicles at a cost of $4 billion had grown by May 2007 to a requirement for 23,000 vehicles at an estimated $24 billion.[86] In an added irony, the MRAP design criteria are nearly the opposite of the AAN/FCS vision; instead of emphasizing standoff, firepower, deep attack, sensor advantages, and light weight, the first priority of the MRAP design is passive platform survivability in the close fight.[87]

Life after the supplemental? Missions and preferences

Supplemental funding has thus far enabled the Army to support ongoing operations in Iraq and pursue ambitious plans to build a modular force equipped with FCS. While the Iraq commitment is unlikely to end within the next two years, the Army faces a seemingly unavoidable collision between its demands for capital modernization and the politically and experientially mandated increases in ground forces. The time may not be far off when the Army will be forced to choose what it wants to retain: a five-brigade increase in end strength, a recapitalized legacy force, specialized equipment for operations in Iraq, the FCS, or modular redesign. In the absence of lavish supplemental funding, it is hard to imagine that the Army will be able to afford all five of its "first priorities."

If the post-Vietnam experience is any indication, the Army's longstanding preference for conventional war will survive the Iraq experience. The vision of future war developed under the AAN project and embodied in the FCS and modularity initiatives was an expression of the institution's preferences more than it was a rational estimate of future war. Ironically, the objective contribution of wartime experience, particularly after 1999, was the least significant input into the Army's interwar modernization equation.

At the same time, the demand for soldiers by policymakers will most likely remain focused on the lower end of the conflict spectrum. Regardless of the outcome in Iraq, some mix of counterinsurgency and constabulary functions will be likely to dominate policymakers' interest in the Army. For every six-week, high-intensity campaign, the Army can reasonably expect a multi-year, labor-intensive stabilization mission. While it is clear that the Army would prefer to fight short, decisive, conventional campaigns, neither the political leadership nor our enemies are likely to accommodate that wish.

Army interwar behavior and theories of innovation

Neither technological opportunity nor experience provides a simple explanation for the course of the Army's modernization in the second interwar period. The technological possibilities of the RMA piqued the interest of Army leaders in the early 1990s but did not immediately provoke sweeping changes in modernization plans. Unsure of the application of the stealth, C4ISR, and precision guidance to land combat, General Sullivan and others chose pragmatic experimentation over radical breaks with the successful Cold War/Desert Storm model. While the Army leadership entertained the possibility of more aggressive technological change, these efforts were limited to think tank studies like the Army After Next (AAN) project.

The greatest influence of RMA technology on Army modernization was indirect. Stealth, precision, and ISR increased the capability and perceived relevance of the Air Force relative to the Army, and it was this threat that did the most to provoke the Army's shift from pragmatic experimentation to revolutionary intoxication. Even this shift did not occur in isolation. Instead, it took the environmental shock of the Kosovo campaign, Task Force Hawk, and resulting civilian recrimination to provoke a strong response from Army leadership.

The influence of environmental signals on transformation varied considerably. While the disappearance of the European Cold War mission injected considerable uncertainty into planning, the Army resisted a shift from warfighting to the constabulary missions of the 1990s. The Task Force Hawk debacle, by contrast, played an outsized role in setting the Army modernization trajectory for the ensuing decade. The Army's apparent inability to contribute to a single, relatively minor regional contingency set in motion radical plans to replace the existing force with Stryker vehicles and later the Future Combat System. By contrast, the multi-year campaigns in Afghanistan and Iraq exerted minimal influence on the Army's modernization plans. Even as close combat in Iraq drew into question the core rationale behind the FCS force – situational awareness and standoff as substitutes for mass and armor – the Army continued to march in the direction of a future force designed for rapid deployment and the deep battle.

Although there is evidence to support elements of all three theories of innovation, interservice competition appears to offer the most compelling explanation of Army behavior during the interwar period. Barry Posen contends that external threats drive civilian intervention and hence military innovation. The mechanisms for such change are civilian sponsorship of maverick military leaders and the distribution of procurement funds. The defining characteristic of the second interwar period was the paucity of serious threats to American survival. As Posen's theory predicts, low threat levels tended to produce civilian indifference rather than intervention. This extended not only to the Clinton administration but also to the first year of the Rumsfeld era. While Rumsfeld and the Army came to blows over procurement projects, these disputes were a competition over status rather than substance. General Shinseki had set the course of Army modernization in 1999 and his later clashes with Rumsfeld did little to disrupt his plans to move toward a medium weight force composed of Stryker and later FCS platforms. When threat levels rose sharply after 9/11, this shift did not prompt effective civilian intervention in Army modernization initiatives.

There is also some evidence to support the Stephen Rosen's vision of long-term, military-led innovation. Both in his official history of Desert Storm and the AAN project, General Scales sought to identify areas in which the Army could make major advances in its ability to wage "firepower intensive, limited war." While his work and those of other Army officers sought to apply emerging technologies to the perceived shortcomings of the force, their vision had a relatively limited influence on modernization until the Task Force Hawk shock. Absent such a shock, it is unclear whether the more radical prescriptions of these reformers would have had a discernable impact on Army procurement and organization.

On balance, Owen Cote and Harvey Sapolsky's theory of interservice competition offers the most compelling explanation for the Army's behavior in the second interwar period.[88] So long as the position of the Army remained relatively secure, the influence of technological change and expected missions did little to alter the course of Army modernization; the service was content to explore line extensions rather than engage in a risky overhaul of proven organizational structures. Even the opportunity to contribute in the early stages of the Kosovo campaign elicited a negative response from an Army leadership eager to avoid another Balkan entanglement. It was only when the Air Force appeared to win the Kosovo campaign and civilian officials openly questioned the future utility of the Army that General Shinseki moved to implement radical transformation based on Scales's AAN vision. Two ironies emerged from this shift. First, the redefinition of the future force pushed the Army from a heavy, close combat mission to one that emphasized standoff and deep battle. This decision moved the Army into more direct competition with the Air Force for prosecution of the deep battle and the Marine Corps on rapid deployment and forced entry. Who would fight the close battle remained unanswered. Second, the fit between Shinseki and Scales's image of future war and the first four campaigns of the twenty-first century was loose at best. Having staked the Army's future to radical and costly transformation, the Army's leadership struggled to explain the validity of a modernization vision that was seemingly at odds with recent combat experience.

From conservatism to revolution to crisis: army transformation in the second interwar period

The story of Army modernization in the interwar period is more complex than it might at first seem. The Army's reaction to the threats and opportunities of the new era was neither entirely passive nor particularly prescient. The Army spent the first ten years defending force structure and readiness; technological experimentation was limited and conservative in scope. After the Kosovo debacle, it was fear of supersession that prompted the Army to lunge into aggressive technological modernization. That modernization program was rooted in the assumption that future wars would be some cross between Desert Storm and Kosovo; they would be conventional in nature but might demand very rapid deployment of combat forces to the contingency area. This vision of future war formed the basis of Shinseki's breathtakingly ambitious plans for technological and organizational transformation of the Army.

Once the Army had placed its bets on "firepower-intensive, limited war," it faced four wars in quick succession. Only one of these, the major combat phase of the Afghan war, bore any resemblance to the Army's vision. Major combat operations in Iraq, and the counterinsurgencies in Iraq and Afghanistan, featured an entirely different set of challenges and demands. Technological advances did not deliver the anticipated benefits in situational awareness, and the binding constraint on Army effectiveness proved to be the availability of skilled manpower rather than the perceived deficit in capital investment. In spite of these fundamental challenges to the modernization agenda, the Army continued to argue that it could support two counterinsurgencies and a modernization program that included FCS, modularity, and major increases in end strength. Only the end of supplemental funding will force painful choices on mission focus and investment priorities.

Notes

1 Major General Robert H. Scales, Jr., *Yellow Smoke: The Future of Land Warfare for America's Military* (Lanham, MD: Rowman & Littlefield, 2003), p. xi.

2 Orr Kelly, *King of the Killing Zone: The Story of the M-1 America's Super Tank* (New York: W. W. Norton, 1989), pp. 88; Brigadier General Robert H. Scales, *Certain Victory: The US Army in the Gulf War* (Washington, DC: Office of the Chief of Staff, United States Army, 1993), pp. 19–20.

3 The Carter/Reagan buildup also included a number of other ground platforms that complemented the "Big Five," including the HUMVEE light utility vehicle, the MLRS rocket system, and the HEMTT heavy utility truck series.

4 Scales, *Certain Victory*, pp. 13–14.

5 Ibid., p. 12; Douglas A. Macgregor, *Breaking the Phalanx: A New Design for Landpower in the 21st Century* (Westport, CN: Praeger, 1997), p. 157.

6 Actual reductions over the four years following Desert Storm: 760,000 to 495,000 (~35 percent) (Macgregor, p. 21). The Army's pre-Desert Shield expectations of the coming force reduction were nearly the same (see Captain Richard Hyde, "A New Force Structure," *Military Review* (November 1990), pp. 11–18).

7 Direct combat losses as cited in Scales, *Certain Victory*, p. 383. The total number of fratricide incidents in Desert Storm was twenty-eight, with a total loss of twenty-four dead and fifty-seven wounded (source: CDR Danny Waterman, "Fratricide: Incorporating Desert Storm Lessons Learned," Naval War College Thesis, March 5, 1997, p. 8).

8 Major General Barry McCaffrey, the commander of the 24th Infantry Division in Desert Storm, spoke for many of the senior leadership when he argued that, "This war didn't take 100 hours to win, it took 15 years" (source: Scales, *Certain Victory*, p. 35).

9 Macgregor, *Breaking the Phalanx*, p. 19.

10 Ibid., p. 48. While many Army futurists embraced this technological explanation of incomplete victory, there is a strong argument to be made that this interpretation is deeply flawed. Different decisions at critical junctures in the land campaign might well have delivered a more decisive outcome using the same tools. The Marine push from the south drove the Iraqi forces north faster than senior commanders had anticipated; a more measured advance might have given the Army time to close the route of escape near Basra before the Republican Guard had time to withdraw north. More important, the political decision to suspend offensive operations appears to have played the most significant role in permitting the escape of these visions. Had the President and his commanders simply continued their operations for an additional twenty-four to forty-eight hours, the Iraqi forces might well have been destroyed in Kuwait and southern Iraq.

11 General Gordon Sullivan, "Letter to the Army's General Officers, *Force XXI*, 5 March 1993," as published in Gordon R. Sullivan, *The Collected Works (1991–1995) of the 32nd Chief of Staff United States Army, Gordon R. Sullivan* (Washington, DC: US Army Department, 1996), pp. 316–317.

12 General Gordon Sullivan, "Maintaining Momentum While Accommodating Change," Army Green Book, October 1991, as published in Sullivan, *The Collected Works (1991–1995)*, p. 65.

13 General Gordon Sullivan, "Speech to Union League of Philadelphia, October 10, 1991," as cited in Sullivan, *The Collected Works, 1991–1995*, p. 11.
14 Task Force Smith was composed of units pulled from occupation duty in Japan. Their low levels of readiness for conventional combat contributed to the disintegration of the unit during its first engagement with the tank units of the North Korean army.
15 "Sixty Years of Reorganizing for Combat: A Historical Trend Analysis," CSI Report No. 14, December 1999, pp. 58–59.
16 Paul Bracken's 1993 *Washington Quarterly* article "The Military after Next" is generally considered to have been the intellectual spark for the AAN project. Paul Bracken, "The Military after Next," *The Washington Quarterly*, Vol. 16, No. 4 (Autumn 1993), pp. 157–174.
17 Scales, *Yellow Smoke*, p. 5.
18 Ibid., p. xi.
19 Major General Robert H. Scales, *Future Wars Anthology* (Carlisle, PA: US Army War College, 1999), p. xiv.
20 Major General Robert Scales and Dr. John Parmentola, "The Army After Next: Intertwining Military Art, Science, and Technology out to the Year 2025," *Army RD&A*, May–June 1998, p. 5.
21 This term is but one of many employed since the early 1990s to describe the use of aircraft to transport mechanized units within a theater of operations: air mechanization, vertical mounted maneuver, etc.
22 Scales and Parmentola, "The Army After Next," p. 5.
23 In the context of ground forces, stealth refers to the reduction of the full range of signatures. While the reduction of radar cross section is the chief concern in air combat, thermal and visual signatures are generally more important in the cluttered environment of land combat.
24 Scales, *Yellow Smoke*, p. 10.
25 Douglas Macgregor, another leading advocate of modernization, makes a similar point in *Breaking the Phalanx* (p. 49) about the disappearance of "movement to contact" as the primary means of finding and defeating the enemy:

> Aided by enhanced surveillance capabilities in the form of UAVs, airborne radars, and satellites, fewer armored and air-mobile ground forces can now concentrate the effects of combat power against the enemy. Rather than move to contact, "all arms" units electronically search and then destroy the enemy on the battlefield.

26 Scales, *Yellow Smoke*, p. 16.
27 Ibid., pp. xi–xii.
28 Ibid., p. xii.
29 Douglas Macgregor makes essentially the same point:

> General John Shalikashvili, Chairman of the Joint Chiefs of Staff, has said that the "main purpose of the armed forces is to fight and win the nation's wars." It is also clear that the armed forces will have new expanded and diverse mission in an unpredictable, rapidly changing world environment. Not every aspect of an organization for combat will fit well with the future environment. Still, it is the warfighting orientation which must shape organizational change, not operations other than war. In this connection, the question for the Army is whether the division structure is the appropriate combat formation to integrate, exploit, and effectively employ the multitude of emerging military capabilities in the future warfighting environment.
> (Macgregor, *Breaking the Phalanx*, pp. 61–62)

30 Thomas Hargrove, "Bridge-Building Unit's Call-Up Shows Need To Cross Euphrates," *Scripps Howard News Service*, January 16, 2003.
31 General Wesley K. Clark, *Waging Modern War: Bosnia, Kosovo, and the Future of Combat* (New York, NY: Public Affairs, 2001), p. 271; Daniel Byman and Matthew

Waxman, "Kosovo and the Great Airpower Debate," *International Security*, Vol. 24, No. 4 (Spring 2000), pp. 15–16.

32 Clark, *Waging Modern War*, pp. 198, 216.

33 The idea of using Apaches to complement fixed-wing air operations apparently grew out of conversations between General Clark and Admiral Ellis that followed a live-fire exercise in Germany in March 1999 (Source: Bruce Nardulli, Walter Perry, Bruce Pirnie, John Gordon, John McGinn, *Disjointed War: Military Operations in Kosovo, 1999* (Santa Monica, CA: RAND Corporation, 2002), MR_1406-A, pp. 61–62).

34 Clark, *Waging Modern War*, p. 231.

35 Ibid., p. 280.

36 Ibid., p. 198.

37 Ibid., p. 301.

38 Ibid., p. 284.

39 Ibid., p. 302.

40 Ibid., pp. 301–302.

41 Ibid., pp. 305, 337; RAND (MR 1406-A), pp. 63–64.

42 Clark, *Waging Modern War*, p. 289; Dana Priest, "Risks and Restraint: Why the Apaches Never Flew in Kosovo," *The Washington Post*, A1, December 29, 1999.

43 Clark, *Waging Modern War*, pp. 319–320.

44 These forces would later include fourteen M1 Abrams tanks and forty-two M2 Bradley Fighting Vehicles. These platforms were not the nucleus of an invasion force, but simply a security force for the twenty-four Apache helicopters and the two batteries of ATACMS/MLRS launchers (source: RAND (MR 1406-A), pp. 79–80; Dana Priest, pp. A1). For the change in requirements, see also Clark, *Waging Modern War*, p. 277.

45 Clark, *Waging Modern War*, p. 331.

46 RAND (MR 1406-A), p. 81.

47 Clark, *Waging Modern War*, p. 367.

48 General Clark has made two arguments in favor of the contribution of Task Force Hawk and NATO ground forces to the success of the Kosovo campaign. First, he points out that the counter-battery radars assigned to Task Force Hawk were used to direct fixed-wing air strikes on Serb artillery positions (Clark, pp. 337–338). Second, Clark argues that the threat of ground operations amplified the direct effects of the ground campaign (Clark, p. 405).

49 Subsequent experience in Operation Iraqi Freedom gave credence to many of the concerns about Apache survivability in the deep attack role. For an account of the failed raid by the 11th Attack Helicopter Regiment in Operation Iraqi Freedom, see Michael Gordon and Bernard Trainor, *Cobra II: The Inside Story of the Invasion and Occupation of Iraq* (New York, NY: Pantheon Books, 2006), pp. 280–281.

50 As cited in David Jablonsky, "Army Transformation: A Tale of Two Doctrines," *Parameters*, Autumn 2001, p. 45.

51 This account is at odds with the popular interpretation of Secretary Rumsfeld's impact on Army transformation. While there was a violent confrontation between Army leaders and Rumsfeld in the opening phases of the Bush Administration, that clash was over power and control rather than the content of technological transformation. Shinseki's plan for transformation along AAN lines predated the Bush administration. The showdown between Shinseki and Rumsfeld did not materially change the course of that transformation strategy.

52 General Eric K. Shinseki, "Address to the Eisenhower Luncheon, 45th Annual Meeting of the Association of the United States Army," October 12, 1999, p. 3.

53 "Strategic Responsiveness: New Paradigm for a Transformed Army," online, available at: www.ausa.org/webpub/DeptILW.nsf/byid/KGRG-6EGLXE, (last accessed February 10, 2007), p. 3.

54 Thomas Ricks, *Fiasco: The American Military Adventure in Iraq* (New York, NY: The Penguin Press, 2006), pp. 68–69.

55 "Fielding of Army's Stryker Vehicles Well Under Way but Expectations for Their Transportability by C-130 Need to Be Clarified," GAO-04–925, August 2004, p. 11.
56 "Initial Impressions Report, Operations in Mosul Iraq, Stryker Brigade Combat Team 1, 3rd Brigade, 2nd Infantry," Fort Leavenworth Center for Army Lessons Learned, December 21, 2004.
57 GAO-04–925, p. 26.
58 Ibid., pp. 22–23.
59 APS detect incoming threats and defeat them before they reach the platform. The Israeli Trophy system is an example of a contemporary APS system. For a more detailed discussion of FCS and APS, see "Analysis of Processes Used to Evaluate Active Protection Systems," GAO-07–759, June 2007.
60 Slat armor is designed to exploit a specific technical vulnerability of the RPG-7 warhead, and is only effective against relatively primitive variants of that weapon; it offers no protection against advanced RPG variants or other light anti-tank weapons (source: R. M. Orgorkiewicz, "Shifting Focus: Armoured Vehicle Protection Adapts to New Threats," *International Defence Review*, February 1, 2007).
61 "Initial Impressions Report, Operations in Mosul Iraq," pp. 49–51.
62 R. M. Ogorkiewicz, "Worldwide Wheeled Armoured Vehicle Programs Raise Questions over Mobility," *International Defence Review* (January 2000).
63 Scales, *Yellow Smoke*, p. 16.
64 The 2006 war in Lebanon is another case of irregular opponents proving the limits of combined arms operations. In Lebanon and Iraq, Arab opponents who elected to confront modern Western forces on symmetric terms lost, while those who pursued irregular strategies were far more successful. Given these results, it would be reasonable to assume that future opponents will pursue irregular strategies until first world armies devise compelling answers to this challenge.
65 *The Army's Future Combat Systems Program and Alternatives*, Congressional Budget Office (CBO), August 2006, pp. xii.
66 Government Accountabilty Office, "Future Combat System Risks Underscore the Importance of Oversight," GAO-07–672T, March 2007, pp. 1–3.
67 The cost figures in this series are presented in 2007 dollars. The original program cost estimate in 2003 dollars was $80 billion. The Army's official estimate of program cost is $160.9 billion. The estimate of between $203 and $234 billion is drawn from an independent appraisal by the Office of the Secretary of Defense (source: "2009 Is a Critical Juncture for the Army's Future Combat System," GAO-08–408, pp. 41; ibid., GAO-07–672T, p. 7).
68 CBO, *The Army's Future Combat Systems*, pp. 41.
69 Ibid., p. 30.
70 Ibid., p. 40.
71 Almost from the start, the FCS program has broken with established best practices for defense procurement by entering product development with a number of immature technologies:

> The business case for a program following best practices in a knowledge-based approach is to have all of its critical technologies mature to TRL 7 (fully functional prototype in an operational environment) at the start of product development. For the FCS, this would mean having had all technologies at TRL 7 by May 2003. By comparison, even with the progress the program has made in the last year, fewer than 35 of FCS's 46 technologies have attained a lower maturity – TRL 6 – three and a half years after starting product development. Immature technologies are markers for future cost growth.
>
> (GAO-07–376, p. 21.)

For similar commentary on the high technical risks of FCS, see CBO, pp. 34–35.
72 CBO, p. 37.

73 Peter Wilson, John Gordon IV, and David Johnson, "An Alternative Future Force: Building a Better Army," *Parameters* (Winter 2003–2004), p. 21.
74 CBO, pp. xviii–xix.
75 CBO, p. 32.
76 Ibid., GAO-08–408, p. 16.
77 This observation was closely related to the TF Hawk experience in Kosovo. For detailed justification of this, see RAND/Arroyo report on Army in Kosovo (MR1406, pp. 129–130).
78 Andrew Feickert, "US Army's Modular Redesign: Issues for Congress," CRS-RL32476, May 2006, p. 2; Government Accountability Office, "Army needs to Provide DOD and Congress more Visibility Regarding Modular Force Capabilities and Implementation Plans," GAO-06–745, September 2006, p. 9.
79 Elaine M. Grossman, "Study Finds Army Transformation Plan Weakens Combat Capability," *Inside the Pentagon*, January 26, 2006, pp. 1–4.
80 Andrew Feickert, "US Army's Modular Redesign: Issues for Congress," Congressional Research Service, May 5, 2006, p. CRS-10; CBO, p. 32.
81 Feickert, "US Army's Modular Redesign: Issues for Congress."
82 "General: Bigger Army Aimed at Keeping US Edge in Conventional War," *Inside the Pentagon*, January 25, 2007, pp. 1–2.
83 Lawrence Korb, Loren Thompson, and Caroline Wadhams, "Combat Fatigue: Vehicle Wear and Tear in Iraq Jeopardizes US Army Readiness," *Armed Forces Journal*, June 2006; Gerry Gilmore, "Army's Equipment 'Reset' Program ahead of 2006 Pace," *Armed Forces Press Service*, February 1, 2007.
84 Greg Jaffe, "Despite Its $168 Billion Budget, The Army Faces a Cash Crunch," *Wall Street Journal*, December 11, 2006, p. 1.
85 "Army Details $10.2 billion in 'Unfunded Requirements' for 2008," *Inside Defense*, February 12, 2007.
86 "Giambastiani, Vice Chiefs to View Final Rounds of MRAP," *Inside Defense*, May 17, 2007; "Army Eyes $10 Billion in Procurement Cuts to Fund Larger MRAP Fleet," *Inside the Pentagon*, May 17, 2007.
87 As the program manager of the MRAP program explained: "This program has two principles… Principle one, you must have a survivable vehicle; the second: they must meet our schedule and be ready to produce" ("Giambastiani," *Inside Defense*).
88 Owen Cote, Jr., *The Politics of Innovative Military Doctrine: The United States Navy and Fleet Ballistic Missiles* (Dissertation, Massachusetts Institute of Technology, 1993); Harvey Sapolsky, "On the Theory of Military Innovation," *Breakthroughs* (Spring 2000).

4 The Navy after the Cold War

Progress without revolution

Benjamin H. Friedman

Contrary to the hopes of advocates of the Revolution in Military Affairs (RMA), the US Navy has seen little innovation since the Cold War. Despite considerable change, the Navy is basically a smaller and no less expensive version of what the United States bought during the Cold War. This chapter demonstrates this point and explains its cause.[1]

The hypothesis behind the so-called revolution is that technological advances in precision munitions, sensor technology, information technology, and communication systems have the potential to revolutionize warfare and the organizations that make it, including the Navy. Theorists of this so-called revolution rarely state what exact shape this revolutionized Navy should take, making it difficult to measure progress. But we can use academic literature dealing with organizations to define what constitutes innovation in a military organization and then ask whether the transformation in the US Navy meets that definition. Because the answer is mainly no, transformation has failed to achieve its advocates' revolutionary ambitions.

In the post-Cold War Navy, there has been plentiful progress as a result of the technologies detailed in Chapter 2, and several significant organizational shifts, but little innovation. Despite considerable change – upgraded and retired platforms, new communications systems – the Navy largely retains its Cold War mix of missions within its three dominant communities: carrier-based aviation, surface ships, and submarines. Change has preserved the utility of platforms by adjusting their missions but not replacing them. The current procurement plan, despite the inclusion of littoral combat ships, offers little evidence of coming revolution.

There are several explanations for the Navy's stasis, and that result is probably over-determined. First, the technologies associated with transformation served existing Navy platforms and doctrine. Second, as organization theory predicts, the Navy was open to change that served existing purposes but not to new purposes.[2] Third, no state threatened US military hegemony, frightening leaders into doctrinal upheaval.

The international environment did affect the Navy, however. A benign security environment lowered defense budgets and made the Navy smaller. The fact that the remaining enemies had small or nonexistent navies threatened the Navy's relevance and sent it looking for a new strategic justification, which affected force structure. But these changes were an evolutionary reaction to international politics rather than a revolutionary embrace of new technologies.

Naval transformation and military innovation

Authors who write about the revolution in military affairs generally fail to state whether the changes they envision are inevitable outcomes of technology or whether they require political action.[3] It is fair to proceed, however, as if the literature is saying what is desirable given technological progress and that it must be produced by political means. This view imposes artificial order on writings that fly under the transformation flag, but it informs this chapter. Anyway, few still embrace the former, technologically deterministic view. It no longer seems plausible, given the continuity in the US military over the past fifteen years. In most cases, some political process, whether driven by dynamics within the state or a strategic problem from without, is needed to make potential innovations real.

The main idea behind naval transformation is that naval surface ships, submarines, and aircraft will become nodes in a network that shares information among its components at high speeds. The advance of sensor technologies, mostly radars on ships, aircraft, and in space, along with space-based infrared (IR), electronic intelligence gathering (ELINT), and the global positioning system (GPS), gives the Navy access to extensive and timely information about the location and characteristics of objects on the ocean and even land, whether friend or foe. Advances in information processing and communications systems allow this information to be rapidly communicated among dispersed elements, processed by computers for targeting, and used by commanders at far greater speeds than in the past.[4] Massive improvements in weapons' accuracy – whether ship- or submarine-launched cruise missiles, aircraft-launched missiles, or laser-guided or GPS-guided bombs – allow commanders to use this rapidly processed and distributed information to target enemies quickly and precisely. The added range provided by cruise missiles and aircraft stretches the area precision fires can target. Stealth technology allows aircraft and, to a far lesser extent, ships, to avoid enemy sensors as they detect and target enemy platforms. Because ships and other platforms coordinate with others to track and strike enemies, we are now in an age, some say, of network-centric warfare.[5] Different elements of a network can collect information, build a common picture, and pass information to the element of the network that attacks the enemy with precision.

As Admiral William Owens, one of the principal exponents of the RMA puts it, the wealth of information that commanders can now use may lift the "fog of war," Clausewitz's description of the uncertainty that distance and time impose on battle. In the past, military commanders used mass to overcome this uncertainty. That is, because commanders were perpetually unsure of both sides' location and actions in battle, they were well-served to send large numbers of forces with plentiful firepower to the enemy's vicinity. War was about applying mass. Now, RMA exponents say, it is about information that lifts Clausewitzian fog. Network-centric warfare allows dispersed elements to deliver massed effects against enemies. This means that when naval commanders are supplied with information about enemy ships, aircraft, and even land assets, they can make quick decisions and instruct the best-positioned asset to strike the enemy precisely without gathering ships together to mass firepower and defenses.

That, in a nutshell, is the theory of transformation as applied to the Navy. The next question is how to measure what progress has been made toward this vision. What changes in forces and doctrine could we expect if transformation occurs? One can measure progress both by the aims of its advocates and, as mentioned, by reference to more academic innovation theory. The failure of transformation literature to translate the ideas just articulated into prescriptions about organization, budgets, and procurement makes the first method difficult. On some implications, however, transformation theorists mostly agree.

One such implication is decentralization.[6] If dispersed elements can cooperate and achieve massed effects, the organizations containing them can become flatter and more decentralized with decision-making distributed to lower organizational levels, it is said.[7] This notion that transformation should include the decentralization of authority is questionable. The US Navy always sought to use available means, from radio to radar to GPS, to centrally coordinate its elements. What we call transformation merely accelerated this process. Better communication makes decentralized execution easier, but it also gives far-off commanders greater awareness of the battle and greater ability to direct it. Transformation then militates against decentralized authority. The argument that authority should be more widely distributed might be right, but it hardly follows from the technology of transformation.

Many advocates of transformation also claim that one of its consequences is smaller ships. This is perhaps another manifestation of the rise of information over mass. Information advantages and stealth technologies might allow US forces to target enemies that do not see them or at least cannot bring their full firepower to bear. On the ground, the notion is that lighter, more vulnerable forces will be able to defeat heavier, more mechanized forces by avoiding meeting engagements – that is, by targeting them before coming into range of sight. At sea, the idea is that smaller ships will be able to use sensor-provided information to engage better-armed enemies that are blind to their presence. This situation might occur as a navy-on-navy engagement, but it is also imagined as part of a coastal assault where small ships brave shore-based munitions. Advocates see swarms of light boats attacking enemies in lieu of larger formations of heavier, more costly ships. Thus some naval transformation advocates called for small, cheap ships called Streetfighters, meant to fight in littorals and take risks that larger, more expensive ships would not.[8] The legacy of Streetfighter is sometimes said to reside in the Littoral Combat Ship (LCS), a frigate-sized ship that is only small relative to the rest of the Navy.

Transformation is also supposed to make the military cheaper. Owens claims that transformation is necessary for precisely this reason. Writing in the late 1990s, he feared military requirements that exceeded military budgets and advocated transformation as a way to meet or exceed requirements within those budgets.[9] The logic here is that network-centric warfare economizes the use of the force, allowing you to do more with less. Cheaper naval budgets that buy equal capability might then be a measure of transformation. Because defense budgets have been expanding while transformation is theoretically implemented, the success of this measure is hard to gauge. Moreover, if transformation allows military forces to do the same or more for less, they can also provide more bang for more buck. Transformation might then allow services to ask for and promise more. Looking at the top line of military budgets to assess transformation's progress is then unhelpful. Looking at the cost of destroying a target would be a better measure, but a hard one to make.

Another widely mentioned outcome of transformation is that the military should become more unified. Enabled by communications systems, the Navy should be able to share intelligence and coordinate its operations, particularly aviation, with its sister services.[10] Jointness has become something of a secular religion within the US military since Goldwater-Nichols in 1986, so it is no surprise that the new sect of transformation should invoke it.

Other advocates of transformation call for greater financial commitment to research and experimentation in the Navy at the expense of procuring large platforms.[11] Their view is that the best way to employ the transformed Navy is not yet clear, so extensive experiments and technological creativity are needed to find it. Heavy spending on research and development would take funds from expensive platform procurement programs favored by the Navy's main communities, transformation's advocates say. Increased spending on research, development and experimentation, however, can only be seen as slight evidence of transformation. There is a difference between innovation and increased efforts to create it.

Finally, transformation's advocates call for vast increases in the amount of money spent on the technologies that make information networks. They want more funds going to the intelligence, surveillance, and reconnaissance (ISR) technology and IT systems that gather and process the information that has become so vital to US military operations.[12]

Using transformation literature to assess the progress of transformation in the Navy, we have the following set of indicators of transformation: lower cost per target destroyed, increased jointness, and a higher percentage of naval spending going to smaller ships, ISR platforms, computing, and communications systems. Lower total military spending and decentralized authority, although sometimes claimed as transformative outcomes, are not considered such here.

The decline of the Soviet Union and failure of other navies to challenge US naval hegemony might cause these outcomes, rather than the transformative technologies. The imperative for joint operations comes in part because the United States so overmatches its opponents that the Navy, rather than fend off the opponents' navies, can dive right into the main fight alongside the other services. Because we plan wars on weak states without navies, the Navy seeks relevance by going into littoral, brown water areas with small ships better suited to that environment than more expensive, larger ships. Increased experiments and investment in research and development might be the result of uncertainty about how to best employ the fleet given the dearth of enemies. If peace, not technology, caused these changes, it does not mean transformation has not occurred. But we should be careful not to give too much credit to technology, as opposed to international politics.

Because transformation literature does not give us an easy way to judge its success, better to measure transformation's progress by its success in causing innovation as defined by students of military innovation.[13] By these lights, innovation entails not only the creation of new ways to achieve missions – new doctrine – but the destruction of a function rendered irrelevant.[14] One doctrine, one way of achieving an end, replaces another. Budgets, promotions, and procurement reflect the change. Destruction need not be official, as the US military often finds new uses for old technologies, or keeps archaic functions intact with far less funding. For example, the most frequently noted naval innovation in the last century is the advent of aircraft carriers during World War II. Prior to the war, the dominant doctrine in the Navy called for battleships to use guns to destroy enemy ships. During the war, carrier-based aviation displaced battleships as the principle offensive tool of the Navy. Surface ships became protectors and thus servants of the carriers.[15] These platforms lost their primary mission and found a new one. Budgets and promotion reflected the change as aviators rose in the naval hierarchy. Innovation can eliminate missions and hierarchies without eliminating ships.

One might contend that these power shifts are too much to ask of transformation. The Navy constantly shifts its priorities and upgrades its technology. Many find it confusing to hear that such change is not innovative. Transformation should, however, mean something more than the updating of military technologies that is common to military services throughout history. Otherwise transformation is a synonym for change and means little. The concept stems from the "revolution in military affairs," after all, a phrase not lacking in ambition. That is why the destructive aspect of innovation is essential. Innovation requires organizational pain, retraining, and reorienting forces, employing new doctrine.[16] It creates losers, as well as winners, which is why it is often opposed. So when we look for evidence of transformation's success in the Navy, we do not look for evolutionary changes that improve the execution of existing tasks, however effective. We seek evidence of the disruption of a doctrine in favor of a new one.

Naval politics

The Navy is the most catholic of the US military services. Where the Air Force has a dominant ethos of strategic airpower, the Marine Corps an adherence to expeditionary actions and amphibious assaults, and the Army a focus on conventional ground combat, the Navy has performed a variety of functions in its history. Its missions today include providing sea control via the suppression of other navies; striking the shore with guns, missiles, Marines, and aircraft; anti-submarine warfare (ASW); mine-clearing; missile defense; presence; and humanitarian relief. Historically, the Navy's missions have also included commerce raiding, blockades, gunboat diplomacy, coastal defense, and amphibious assaults. The Navy does not have as fixed an identity as the Army or Air Force.

The Navy is more like a confederacy to the other services' federal unions. It includes three primary communities: surface warfare, submarines, and aviation. Each community has a high degree of independence. Each is a constituency within the Navy for performing certain missions and buying associated platforms. The Marines (not treated much in this chapter because they are discussed in Chapter 6) often operate in amphibious-ready groups, which ride amphibious assault ships, amphibious transport docks, and dock landing ships.[17] Surface warfare includes major surface combatants: frigates, destroyers, and cruisers. Aviation includes carriers and their associated air wings, which today consist of fighters, electronic warfare and surveillance platforms, cargo planes, and various helicopter types.[18] Submarines include ballistic missiles submarines, fourteen of which carry nuclear-tipped ballistic missiles, and attack submarines, which can attack surface ships, fire cruise missiles, conduct ASW, and collect intelligence.

The communities themselves do not have methods as fixed as those of the other services. In American history, destroyers, for instance, have protected battleships, hunted submarines, supported amphibious assaults, and fired missiles at distant shore targets, among other roles. They have operated in a variety of groups and task forces working with changing mixes of Marines, other services, and allies. Their crews train for most of those missions today.

The missions of the three communities overlap. Fleet defense is provided by air and surface forces. Submarines, aircraft, and surface ships participate in antisubmarine warfare. All communities can strike sea or shore targets. The Navy's various missions and communities give it various cultures and doctrines.

The Navy's medium, the sea, and its oldest vehicle, the ship, heighten its penchant for autonomy and diversity. Unlike the Air Force, the Navy matured in an age where technology offered little control of dispersed platforms, particularly ships. In a vast ocean, before satellites, beyond the reach of radios and telegraph, phone, and supply lines, the naval captain received orders in port, perhaps, but then became the king of his domain. The autonomy born of this condition is captured in naval culture even though communication technologies have destroyed its original cause. This was especially true in the early nineteenth century when the Navy was more interested in gunboat diplomacy than the sort of global dominance that required the defeat of rival navies and thus the concentration of battleships.[19] Ships in formation coordinated action by training, sight, and later radios, vesting more authority in the battle group commander. But relative to an Army unit, command was dispersed. When submarines joined the fleet, they continued this tradition of autonomy. Their missions often required working alone and, being underwater, they were especially remote from communications networks.

The Navy's decentralization has two relevant consequences. First, it is a potential source of innovation that the other services lack. Diversity of methods in service of overlapping missions means that the communities compete. Competition provides pressure to innovate and gain funds and stature at the expense of the other communities.[20]

Second, in part because of its penchant for distributed authority, the Navy has traditionally resisted service unification. It fought the creation of the Department of Defense after World War II. It initially fought Goldwater-Nichols, which furthered the integration of the services in 1986. Today the Navy embraces jointness, but perhaps not with the vigor of its sister services.

Ships drive naval politics. They must be purchased, maintained, and replaced at public expense. Surface combatants, submarines, and carriers last thirty to forty years, longer than the platforms employed by the other services – with the possible exception of Air Force bomber and tankers. Once a ship has been built, it is almost certainly going to be used, given the expense. This makes changing the Navy a very slow process. Today's Navy was largely purchased decades ago.

Ships are also expensive and few in number relative to other services' platforms. They require large annual allocations from Congress and specialized production facilities – shipyards – for construction and upkeep. There are now six major and several minor American shipyards. They provide one of the largest sources of employment in several states, including Mississippi, Virginia, Connecticut, California, Louisiana, Maine, and Wisconsin. Congressmen and senators from these states strive to keep the yards busy. They pressure the Department of Defense to buy more ships and make arguments about why more ships are needed (the rise of China, for example). These representatives sit on the budget committees and pay far more attention to these issues than their colleagues. The contractors that own the yards lobby for their share of the work, as do the unions that represent the workers. These forces both insure the Navy a base of support for its procurement and obstruct any changes – such as the elimination or severe downgrade of one class of ships – that might threaten a yard.[21]

These pressures drive the shape of the Navy as much as international politics. Changes that cut a yard's work will produce resistance among Navy boosters; changes that threaten to close a yard will produce an uprising. Changes in doctrine and especially procurement that threaten a naval sub-community's sense of mission would upset the political structure that runs the Navy. Improving yesterday's Navy is always easier than re-imagining it, because today's Navy reflects a winning coalition, which brought external support and internal harmony. This is not to say that these forces are inviolate, especially during war, but that they are hard to overcome.

The Navy often produces strategy documents, which purport to guide its constitution. But strategies can only change the Navy on the margins of these political realities. The real strategy of the Navy is a product of these political pressures and plays out in budget allocations and procurement schedules. It is there that transformation's success is tested, and there where it comes up short.

1990s

To understand how the Navy evolved after the Cold War, it is useful to consider how it ended that conflict. From a bureaucratic standpoint, the 1980s were boon days for the Navy. The Navy's strategy of the day, labeled the Maritime Strategy, was possible because of Reagan's military buildup. He and his Secretary of Defense, Casper Weinberger, were hawkish, but not attached to any particular use of naval power – they were willing to hand a pile of new money to Secretary of the Navy John Lehman and his service without many strings attached.[22] The Maritime Strategy was predicated on the idea of a global war, where the Navy threatened the Soviet fleet in their northern sanctuaries and tied down Soviet ground forces on the "northern flank." Naval officers loved this strategy because it gave them a more offensive role.[23] The strategy allowed the Navy to approach 600 ships and use its second and third fleets for attack rather than training and replenishment.

During the 1980s, all three naval communities were fairly content. By 1987, the fleet included 568 ships, including carriers, surface ships, and submarines.[24] Spruance-class destroyers and Oliver Perry-class frigates had recently come into the surface fleet. Tomahawk cruise missiles provided range exceeding most carrier aviation, allowing surface combatants to be offensive and do more than protect carriers, in the eyes of their operators. Aegis cruisers, which had far better defense capabilities against aircraft than their predecessors, were in the fleet by mid-decade.[25] The attack submarine community had new Los Angeles-class attack submarines and larger Seawolf-class submarines on the way. Antisubmarine warfare was seemingly going well, with deep-water, long-range submarine detection capability provided by the Sound Surveillance System and an upgrade underway in the early 1980s for the P-3 patrol aircraft that helped track Soviet submarines.[26] In 1987, the United States deployed thirty-six Trident and Poseidon ballistic missile submarines – boomers. Aviators had fifteen carrier decks to take off from in new F/A-18 ABs, and the A-12 attack aircraft, later cancelled, was thought to be on the way.

Things went less well for the Navy after the Cold War, although, in terms of budget, the 1990s drawdown only amounted to giving back the Reagan buildup, not a reduction to historical peacetime levels of ships and spending. The Soviet Union was a middling naval power, focusing its investment on land forces. The Navy justified much of its Cold War force structure based on its contribution to land fights – using carrier-based air power in particular to slow Soviet advances or those by Soviet allies. The Soviets also provided submarines to hunt and nuclear missiles to pre-empt with submarine-launched ballistic missiles. Since 1990, the purpose of the Navy's force structure has been less certain. True, a strategy that involves clearing the Straits of Hormuz or the Taiwan Strait requires some of the Cold War force structure, with contributions from all three communities. But nothing now justifies the sort of Navy that the USSR did.

During the first Bush Administration, the Soviet Union collapsed. By 1990, the Democrats who controlled Congress, led by House Armed Committee Chairman Les Aspin and Senate Armed Services Committee Chair Sam Nunn, sought a significant reduction in military spending, a peace dividend. The Bush Administration, led by Chairman of the Joint Chiefs of Staff Colin Powell, sought to control the drawdown. This review produced "the Base Force," a minimum force level for each service.[27] Powell ultimately concluded that it would be too much trouble politically to cut the forces in any but an equal way. Only the Marine

Corps avoided their fair share of cuts (going down to 170,000 marines rather than 150,000 as originally planned). The other services all received the same percentage reduction in budget. This decision to spread the pain equally without regard for strategic relevance would be repeated in the Clinton Administration's Bottom-Up Review.

Like the other services, the Navy initially resisted the Base Force cuts, which amounted to accepting a 450-ship navy rather than a goal of 600, a 20 percent drop in end-strength manpower, a 20 percent cut in attack submarines, and a drop from fifteen carriers to twelve.[28] The Navy's resistance may have come in part because the pre-eminent community in the Navy at the time was the submariners.[29]

Unlike surface combatants and carrier aviation, submariners almost entirely focused on the Soviet Union during the Cold War. To a submarine operator, particularly an ASW person, the end of the Cold War was not the event it was for the rest of the world. The Soviet/Russian missile capability did not change as fast as Soviet politics. Even though Soviet submarines did not patrol as much as during the Cold War, they remained armed with nuclear missiles capable of attacking the United States. It was hard for the submarine fleet to accept that 50 percent fewer submarines were sufficient to match these forces.[30] The Chief Naval Officer, Admiral Carlisle Trost, never agreed with the Base Force cuts. But when he retired in 1990, his replacement, Admiral Frank Kelso, did cooperate. Powell's power, a lack of Congressional support, and events in Europe left little alternative.[31]

The first seminal event of the post-Cold War period for the US military was the Gulf War. Despite the easy American victory, the war was unsettling for the Navy. Although submariners fired cruise missiles and did some intelligence gathering, the war served as a reminder of their diminished relevance in a world where US adversaries were rogue states with small or nonexistent navies. Surface combatants policed the Persian Gulf and fired a considerable amount of Tomahawks, but their contribution was small and expensive relative to the firepower provided by the Army and the Air Force.

The Gulf War left naval aviators with more egg on their faces, for several reasons.[32] First, the Navy had resisted the creation of a new position under Combatant Commander Norman Schwarzkopf, called the Joint Force Air Component Commander (JFACC). The JFACC, here Air Force General Charles Horner, was to coordinate US airpower. This role was a natural outgrowth of the 1986 Goldwater-Nichols legislation, which created the Combatant Commander position to coordinate the services during war. Fearing subordination to the Air Force, the Navy hoped the war would be run like Vietnam, where geographic regions were given to each service for airstrikes. Horner, with the blessing of Schwarzkopf, insisted on an integrated air campaign. Given its weakness on Horner's staff, this left the Navy in a subordinate role in deciding on targets and tactics.[33] After the war, conventional wisdom embraced JFACC as a success that should be implemented in future conflicts. The Navy's initial resistance could be seen as evidence of the wrong-headed parochialism that had caused it to oppose both the creation of the Department of Defense in 1947 and Goldwater-Nichols.

Second, the Navy had not invested in laser-guided bombs (LGBs) to the extent the Air Force had (GPS-guided, Joint Direct Attack Munitions – JDAM – did not yet exist). The Navy decided during the air campaign to marshal its small supply of LGBs for the ground campaign, which was ultimately too brief to use them. The Navy wound up dropping only 7 percent of the war's smart bombs.[34] When the war's aftermath saw these weapons celebrated as the leading symbol of American military prowess, the Air Force received the credit, and naval aviation looked relatively backward.

Third, the Navy E2-Cs, electronic warfare aircraft designed for fleet air defenses, lacked radar allowing them to reliably distinguish aircraft over or near land. As a result, they failed to see two Iraqi Mirages coming down the Iraqi coast, which the Air Force Airborne Warning and Control System (AWACS) picked up. This, too, showed the Air Force's relative sophistication.[35]

Fourth, carriers had to receive the air-tasking order (ATO), which assigns daily targets to strike aircraft, by hand delivery rather than electronically. To many, this was another sign of the Navy's parochialism and second-rate technology. This perception was somewhat unfair, given that the same problem was common to some Air Force units at the time.[36] Nonetheless, the impression had political consequences.[37]

These failings were a consequence of the fact that under the Maritime Strategy, the Navy planned to fight its own air war, with many of its aircraft to be used for fleet air defense over water, while the Air Force and Army fought the Soviets on the Central Front. The crowded skies imagined over Germany would require order to avoid collisions and to make efficient use of resources. That was the job for the ATO, although it could only deal with fixed targets identified well before attack – generally the sort of strategic targets beloved by the Air Force.[38]

The end of the Cold War and the Gulf War compelled the Navy to worry about its role in American defense.[39] The Navy saw a need for a strategy to articulate its relevance to American defense and encourage the ranks. When the Navy embarked on a process from 1991–1992 to map the future, the "Naval Forces Capabilities Planning Effort," they came up with a strategy, "...From the Sea," which emphasized attacking the shores. Mahanist notions of naval rivalry would not do. No rival navy existed to threaten US control of the seas. Essentially, the Navy had to go where the enemies were to stay in business.

The strategy, which came out in 1992, emphasized forward presence in foreign waters and sea strike – meaning aviation, missiles, and amphibious assaults of Marines. The idea is that naval forces can be used quickly and flexibly for a variety of options ashore and that their mere presence offshore stabilizes regions where they patrol. The Navy always notes that they need not ask permission for a base like the Air Force.[40]

The idea of naval missions striking land targets is old. The Navy often engaged in gunboat diplomacy in the nineteenth century. In the later stage of the Pacific campaign in World War II, the Navy mainly supported forces on shore. Naval support for the wars in Korea and Vietnam also came from the sea to land. In the early Cold War, the Navy, with no real Soviet navy to engage, focused on delivering nuclear bombs and missiles to Soviet land targets. Under the Maritime Strategy, the Navy planned to attack land to draw Soviet forces away from the fight in central Europe. What the "...From the Sea" strategy marked was a change of emphasis.

Clinton's Secretary of Navy, John Dalton wanted a strategy document not written under a Republican, however similar to the last strategy.[41] The result was "Forward from the Sea," essentially a 1994 recapitulation of "...From the Sea." The "From the Sea" idea, while not detailed, remained the essential template for the Navy throughout the 1990s. It still does, in a sense, even though the name has changed every few years.[42]

Essentially, the new strategy was a way of justifying the Navy, not a starting point to reinventing the Navy. Thus it is no surprise that these documents did not call for the elimination of mission areas but for slightly different employments of existing platforms. Nonetheless, they reflect a geopolitical reality that affected the Navy's force structure.

If the future was attacking targets from the sea, the failings of the Gulf War would have to be corrected. The Navy needed to get better at flying with the Air Force over land masses and providing support to ground forces. Changes were made to this end. Naval aircraft gained the ability to communicate with Air Force AWACS aircraft, which fly behind the battlespace using powerful radar and communications systems to quarterback the skies. By the time the Kosovo conflict occurred, the Navy had increased its representation in the Combined Air Operations Center (CAOC) run by the Air Component Commander to coordinate airpower. The E2-C's (Hawkeye's) radar was improved, although only the advanced version, the E2-D, which has not entered the force, will have look-down, shoot-down capability over land.[43] The ATO was made to come electronically.[44] The investment in smart bombs, both laser- and satellite-guided, was heavy. F-14s gained an air-to-ground strike capability in 1997 when they were upgraded with the Air Force's LANTIRN (low-altitude navigating and targeting infrared for night) targeting system, which allowed the delivery of laser-guided bombs.[45] F/A-18s gained the ability to drop the satellite-guided GBU-31 2,000-pound JDAM.[46] F-18E/Fs extended the unrefueled strike range from carriers.

After the Gulf War, jointness became not only an operational reality but also a political imperative. Within the Navy, its champion was Admiral Bill Owens, who was Deputy Chief Naval Officer (CNO) for Resources, Warfare Requirements, and Assessments from 1991–1993 and Vice Chairman of the Joint Chiefs of Staff from 1994 to 1996. By the mid-1990s, the Navy had stopped asserting in strategy papers and speeches that it was joint with the Marines and that the Army and Air Force could be joint with each other.

With its emphasis on strike and presence, the strategy privileged the carrier community and, to a lesser extent, surface ships, while limiting the role of submarines. Nonetheless, the end of the Cold War meant that even the winners lost force structure throughout the 1990s. As proposed in the Base Force Review, the carrier force fell from fifteen (with one carrier always in overhaul) to twelve during the 1990s. The total number of surface combatants (cruisers, destroyers, and frigates) fell from 220 in the late 1980s to 125 by 1996.

Aegis fleet defense capability and the Tomahawk cruise missile kept the surface community relatively happy during the 1990s, despite the loss in ships.[47] Ticonderoga-class cruisers accommodate 122 vertical launch tubes for cruise missiles. The Navy continued to deploy new Arleigh Burke destroyers, and planned to deploy larger DD-21 destroyers (later this program became the DDX and DDG-1000 program). Destroyers were tied to the strategy in part by their ability to provide gun and cruise missile support to Marines ashore, either on a beach or further inland – what the Marines bill as "ship-to-objective maneuver."

Another outgrowth of the heightened focused on striking the shore was the arsenal ship. These ships, proposed in 1996 with support from the Defense Advanced Research Projects Agency (DARPA), were designed to provide fire support for operations on land. They were to carry 500 cruise missiles, four times that of a cruiser. They were to be heavily automated with a crew of only fifty, and receive targeting information from cruisers using the Navy's Link 16 and Cooperative Engagement Capability (these communications systems are discussed more below, pp. 85–86). Despite a "stealth design," which would reduce but hardly eliminate radar signature, a double hull and the ability to partially submerge, they were quite vulnerable to attack, due to their large size and slowness.[48] At an estimated $500 million, they cost only a fraction of what a destroyer or cruiser does, although initial price tags rarely hold. The initial plan was to procure one as a concept demonstrator. Had the arsenal ship been built, it might have threatened destroyers' role in providing fire support ashore. According to its advocates, it might even have competed for certain strike missions with B-2 bombers.[49] In this sense, the program had the potential to be innovative.

The Navy never built the ship. It ran aground against the hostility of the surface community and its Congressional backers (largely worried about shipyards), not to mention other naval communities worried that their budgets would pay for the ship.[50] The defense budget for FY 1998 largely eliminated funding for the program. With some merit, the arsenal ship is often portrayed as a casualty of the Navy's hostility to transformation.[51]

Another consequence of the Navy's increased commitment to attack from the sea was the creation of Expeditionary Strike Groups in the early 1990s. These formations are bulked-up versions of amphibious-ready groups. They include a Marine Expeditionary Unit, an amphibious assault ship carrying thirty helicopters and six jump jets for Marines, a dock landing ship, and an amphibious transport dock. ESGs include these features plus a destroyer, cruiser, frigate, attack submarine, and a P-3C Orion aircraft. Currently, the Navy plans to maintain eight, down from the originally planned twelve.[52] It is unclear what realistic purpose the ESGs serve. If the Marines were to mount a hostile landing, there would be almost certainly be carriers – probably four or five – available to support their landing.

Submariners have a diminished role in a naval strategy that focuses on land war. Attack submarines enable strikes from sea by defeating enemy craft more than by performing strikes directly. They are traditionally justified by their contribution to American sea control, largely through ASW, a mission that today faces little challenge outside of waters near land. Cruise missile launch tubes gave attack submarines a strike mission starting in the late 1980s, putting them into competition with surface ships and, to some extent, carrier-based aviation. In recent years, Virginia-class attack submarines also gained the ability to carry SEAL delivery vehicles to deploy special operations forces for shore-based operations, although there have been problems with the mini-submarine delivery vehicles.[53] While submarines have long collected signals intelligence, they have focused more on that mission in the post-Cold War period.[54]

Despite these moves to retain relevance, attack submarines lost much of their raison d'être at the end of the Cold War with the absence of a sea rival.[55] The Pentagon curtailed the Seawolf program at three boats in 1995, and began the design of the smaller Virginia-class attack submarine. The Navy went from ninety-four to fifty-four attack submarines during the 1990s. Today the figure is fifty-three and probably headed lower.[56] The fact that attack submarines now spend the majority of their mission days collecting intelligence, rather than in traditional roles, might be called transformational. But this appears to be a consequence of not having much else to do, rather than a truly innovative function that threatens another community.

Ballistic missile submarines (SSBNs), designed for nuclear deterrence, are even more a Cold War creature. With the conversion of four to a conventional role, their deployed number fell to fourteen. These four are now guided missile submarines (SSGN) armed with cruise missiles (154 each), and they carry up to sixty-six SEALs that can be launched to shore.[57] With these new missions, SSGNs are a potential competitor for aviators, surface ships, and various Air Force platforms. The Navy's current shipbuilding plan does not provide for the replacement of the SSGNs when they retire, although this may change.

The 1993 Bottom-Up Review, led by Les Aspin, who had become Secretary of Defense in the Clinton Administration, followed on the Base Force as an articulation of national force requirements. The aim, which failed, was to rethink the premises of the defense establishment and save money, rather than simply shrinking the existing force as the Base Force had. In the end, the Review simply cut the services equally to avoid resistance. The review proposed a force of eleven carriers plus another so-called training carrier – one less than the Base Force total – 346 ships, including forty-five to fifty attack subs.[58]

The Review theoretically used three missions to arrive at its recommended force structure: two nearly simultaneous major regional contingencies (i.e. the idea that Iraq would attack Kuwait and North Korea attack South Korea within a period of weeks), presence, and peacekeeping. Only the first two missions are relevant to the Navy. The planning scenarios used for the Base Force envision four or five carriers for each major regional contingency, which was probably excessive. A RAND study at the time used three.[59] Major regional contingencies then created a need for only ten carriers. The eleventh carrier was justified by the presence requirement.

The theory behind the presence requirement was that the United States should have forces at all times in three so-called hubs: the Mediterranean, the Indian Ocean/Persian Gulf, and the western Pacific. Given the homeport time crews need, travel time, and repair, eleven carriers could maintain a carrier full-time in one region, and keep carriers in two other regions with gaps of a few months annually. As various authors note, there is little evidence that presence has much effect on local politics or that carriers rather than something smaller and cheaper are necessary to show the flag.[60] It is one thing to send a carrier on a presence mission when its cost is theoretically justified by other missions; another matter to justify the carrier with that mission when a cruiser or an amphibious assault ship might suffice. The Navy argues that forward presence allows quick response, which is true but not necessarily important given that most US military actions are deliberate – involving considerable planning and time. The Navy also claims that the presence of American ships has broad deterrent effect and underlies global trade. These claims are difficult to demonstrate or refute with evidence.

By the end of the 1990s, once the Kosovo conflict was over, the Mediterranean was no longer very important to US military operations. Carriers pass through and conduct exercises en route, but there is little reason to keep them there. As a result, the Navy effectively got rid of its requirement to have forces in the Mediterranean, and adopted two hubs: one in the Arabian Sea/Persian Gulf, one in the east Pacific. This made presence a less demanding requirement. So, in recent years, the Navy has argued for its desired forces more in terms of surging carrier battle groups forward during crises and wars.[61] Presence is usually considered a side benefit that the surge force makes possible.[62]

Geopolitics, more than transformational technologies, changed the Navy in the 1990s. Without major threats, there was pressure from the White House, filtered through the Office of the Secretary of Defense (OSD), to contain costs, even though military budgets started growing in the second half of the Clinton Administration. The Navy used strategy to justify its force structure rather than plan it. Driven by unipolarity to the shores of minor rivals, the Navy's internal dynamics and procurement shifted slightly. Submarines lost some budget and prestige and sought more of a shore strike role to remain relevant. Neither the first Bush Administration nor the Clinton Administration imposed a strategy on the Navy, aligning with one naval community at the expense of others. Left to its own devices, it is no surprise that the Navy generally maintained the traditional log-roll between its three dominant elements, and no major naval function was overthrown.

Since 2001

The second Bush Administration came to power speaking the language of transformation. The new defense secretary, Donald Rumsfeld, forced the services to submit annual transformation roadmaps and created the Office of Force Transformation, under the patron saint of net-centric warfare, Vice Admiral Art Cebrowski, to push the concept. The Administration's focus on transformation was disrupted, however, by September 11. Administration officials are fond of noting that they transformed the military while employing it in Afghanistan and Iraq. But the massive budget boost that followed September 11 undermined the impetus toward change. The rhetorical commitment to the theory was not matched by a clear articulation of what transformation means for procurement or by willingness to expend political capital to cancel non-transformative programs.

In recent years, precision and Cooperative Engagement Capability produced tremendous improvements in striking power and fleet defense. These are the realization of decades of research. The improvements to aviation, particularly the embrace of precision weapons and communications systems allowing real-time information exchange for targeting, made the performance of carrier-based aviation in conventional phases of the wars in Afghanistan and Iraq exemplary. Fleet data links have continued to mature, though no competitor tests them in battle.

Data sharing among naval platforms has improved a great deal in speed and volume in recent years. The days of ships' autonomy are long gone. As noted, naval commanders have always wanted to coordinate their forces and share information about the location of enemy forces. The Japanese used radios to coordinate their fleet and rout the Russians in the Tsushima Strait in 1905. High-frequency radios and satellites tightened the coordination of the American fleet in the 1980s. Transformation increased the rate of change in linking naval elements at sea but did not invent the concept of networking. Transformational communication technologies have evolved to link traditional naval platforms doing traditional missions.

During the Cold War, the Navy developed Link 11 – a computer-to-computer digital radio communications program using high frequency (HF) and ultra-high frequency (UHF) – to communicate and create a common operating picture within the battleground and with NATO allies. In the late Cold War, the Navy upgraded the system to Link 16, which uses the Joint Tactical Information Distribution System (JTIDS; see Chapter 2) and allows jam resistance, improved security, increased data speed, and smaller computers, permitting links increased numbers of network participants, including aircraft. High-speed data links today allow video conferencing. The Secret Internet Protocol Router Network (SIPRNet) links the battle group to the national coded military network via satellite, allowing the exchange of emails and chatroom communications among controllers on various ships, and rapid communication of images and attachments. These upgrades allow not only much heightened fleet coordination and shared targeting information, but, along with GPS, real-time knowledge of allied forces' locations and actions. Commanders have gained greater knowledge of the fleet's position and more control over it.

A key communications development is Cooperative Engagement Capability (CEC), which integrates the radar pictures of Aegis-equipped ships. Tested in 1990s and first fielded in 2001, the system combines continually updated data gathered from distributed ship-based radars. CEC allows a lead ship to designate a particular ship to receive targeting data and engage an enemy aircraft or missile.[63] The system was ultimately extended to non-Aegis ships and aircraft, including E-2 Hawkeyes. This technology vastly improved the capability of the US fleet to destroy hostile airpower and missiles, extending its virtual shield outward to far greater distances, but it has not been needed in recent wars.

These communications systems, and several others, form the basis for what the Navy designated in its 2002 strategy document, "Seapower 21," as FORCEnet. The Navy claims that FORCEnet is a central element of net-centric warfare and the glue that unites its forces and missions.[64] FORCEnet, however, is more a rhetorical bow wrapped around existing capabilities for presentation purposes than any new capabilities.

Communications technology has also given a boost to aviation. Carriers now have a tactical strike coordination module that takes the ATO in email form via satellite and feeds it to the Tomahawk and aviation targeting systems.[65] Strike aircraft are now digitally linked to forward air controllers on the ground, who can send text messages and even images to pilots' heads-up displays, allowing pilots to program in GPS coordinates and destroy battlefield targets as they are detected. Laser designation on the ground also allows destruction of moving targets, provided there is no cloud cover. Mission planning centers aboard carriers can use the Rapid Targeting in Cockpit (RTIC) system to send reconnaissance information, including satellite images, to show targets to aircraft hundreds of miles away over a battlefield.

Recent years have also seen great improvements in the range of naval aviation, and especially in the accuracy of the fires it provides. In the past, naval aircraft offered essentially a limited strike option, constrained by range and inaccuracy to raids against littoral areas. F/A-18Es have 33 percent more fuel capacity than the older C version, significantly more range (most claim 30 to 40 percent), more loiter time, and 25 percent more payload. F-35Cs (Joint Strike Fighters) will have an unrefueled combat radius of 700 to 800 nautical miles when they join the fleet.[66] Still, this is less than the range of the A-6, the carrier-based attack aircraft that first flew in the 1960s.[67] The Navy has also increased carrier's sortie generation rate.[68] These changes enable the Navy to sustain carrier-based strikes, potentially for weeks or months, far inland, assuming there are refueling tanker bases available.

In Operation Enduring Freedom, the Navy sustained a constant air presence over Afghanistan, 600 to 700 nautical miles from the carriers, with fighter/bombers usually refueling twice en route and once again from the Navy's own S3 refuelers before landing. Four carriers were employed at once; one, the Kitty Hawk, for Special Operations Forces staging. Six carriers and their battle groups entered nearby waters over the first six months of battle. But, for the majority of the time, two carriers and their air wings were enough. Air Force bombers flying from Missouri dropped the bulk of the JDAM-equipped bombs, but naval aircraft flew the bulk of the sorties – especially the time-sensitive targeting runs where aircraft loitered and received ground targeting information from forward air controllers.[69]

In 2003, over Iraq, carrier-based aviation again maintained a steady presence over a landmass, striking both pre-planned and time-sensitive targets alongside Air Force fighters and bombers. Because Turkey denied the Air Force the basing to fully employ its tactical aircraft, naval aircraft again provided much of the time-sensitive firepower.[70] Naval aircraft flew strikes off three carriers in the Persian Gulf and two in the Mediterranean.

Along with gains in destructive capability, the increased precision of carrier-based aviation has cost-saving potential. Rather than several sorties being required to destroy one target, as in the pre-precision past, one F/A-18 or F-35 can destroy several targets in one flight. You can do more with less.[71] This could lead to the conclusion that we need less aircraft, less deck space, and therefore fewer carriers to produce the same effects. And indeed, since the advent of precision, we have fewer carriers and fewer combat aircraft on each. Moreover, the initial planned buy of F/A-18 EFs was cut from 1,000 to 462 (it now seems likely to increase to around 500), and the initial planned combined Navy and Marine Corps purchase of F-35s from over 1,000 to 680, and it may drop lower.[72] How realistic the initial goals were is questionable. And while capability is up, total costs are not down. The price of destroying a target with airpower has dropped, probably precipitously, but the taxpayer has not reaped the windfall.

On the other hand, the added capability provided by range and precision permits the services to promise more for the same or a higher price. Range gains put a large portion of the world's landmass and the majority of its people within reach of carrier-based aviation, particularly with refueling. Most of the world has become a littoral, according to the Navy. This means that the Navy can do more, even compete directly with the Air Force's tactical fighters.

Jointness prevents such heresies from being openly stated, but given the results in the last two wars, carrier-based aviation could arguably obviate the need for the Air Force's tactical fighters. It could conduct initial strikes against air defense targets, defeat enemy fighters, support ground forces, and allow Air Force bombers to drops bulk loads of smart bombs unmolested. This outcome is unlikely, but it is the sort of creative destruction that would qualify as innovation. Its mere possibility might spark useful competition between the services, and in the muted F-35/F-22 competition, it arguably has.

The centerpiece of transformation in the Navy is the Littoral Combat Ship, at least according to Admiral Mike Mullen (whose tenure as CNO began in 2005 and ended in late 2007, when he became Chairman of the Joint Chiefs of Staff. The new CNO is Admiral Gary Roughead). LCS is often said to be the outgrowth of Streetfighter, the small, fast ship envisioned by Wayne Hughes and Arthur Cebrowski in the late 1990s, which was to displace only about 400 to 500 tons. Streetfighter was to be expendable, unlike expensive surface ships, and use network technology to swarm into attacks in coastal waters full of mines and in range of shore-based weapons. The Navy ultimately would not buy something so small. There is evidence that Rumsfeld's OSD forced LCS on the Navy as something of a replacement for Streetfighter.[73]

With a draft of only ten feet, LCS is also designed for shallow water missions. However, with a displacement of 2,500 to 3,000 tons, it is not nearly as small as Streetfighter. And because the unit cost has ballooned to $500–600 million a piece, the ship is hardly expendable.[74] LCS has more in common with one of the ships it is replacing, the Oliver Hazard Perry-class frigate (which displaced 4,000 tons), than with Streetfighter.

The Navy plans to buy fifty-five LCS to replace fifty-six ships of three types, twelve Osprey-class coastal mine hunters, which were all retired by late 2007, fourteen Avenger-class mine countermeasures ships, which are stationed in Bahrain and Japan, and used for deeper waters, and thirty Perry-class frigates, which are scheduled to retire around 2015.[75] The Navy's anti-mine capability will reside on destroyers (some of which now carry unmanned remote mine-hunting vehicles) and LCS. The retirement of the Perrys reflects the fact that frigates were envisioned mainly as convoy escorts in the Atlantic during a Cold War clash. This change is odd given the emphasis the Navy's current strategy places on commerce protection and interdiction, tasks for which frigates are best suited.

Its boosters say that LCS's modules make it transformational. Cebrowski believed that the fleet must be in a constant state of becoming, meaning that ships would require flexibility to adjust to changing missions within a thirty- or forty-year lifespan. LCS arguably embodies that idea because it is essentially a hull that can allow a series of different missions. Each ship is to be outfitted with one of three modules: one for ASW, one for mine warfare, and one for surface warfare against small craft. This last module includes a UAV for ISR support. Some Navy planners say that LCS will transport SEALs and provide ISR to the fleet. These capabilities are supposed to make LCS relevant to counter-terrorism.[76]

The Navy's support for LCS may be shallow. Admiral Roughead is enthusiastic, but the program is less popular among naval officers, some of whom refer to it as "Little Crappy Ship." The Navy procured the first ship, LCS-1, in 2005, but a stop-work order was issued on the next ship due to cost overruns. Because the current Navy procurement plan seeks more ships (313) than the shipbuilding budget can afford, given current costs, it is possible that naval leaders, especially the surface community, will try to cut or cancel the LCS buy to defend their favored programs.[77] Given that it allows cancellation of three ship types and threatens other platforms' budget share, procuring even several dozen LCS would be somewhat transformational.

Another organizational change undertaken by the Navy in recent years is the Tactical Aviation Integration Plan, which aims to create efficiencies between Naval and Marine aviation. Four Marine strike fighter squadrons were integrated onto carriers in the 1990s. The plan brings the total to ten and puts three naval strike squadrons at the disposal of Marine Corps units. The plan is said to allow the Navy and the Marines to buy a total of 497 fewer F-35s, operate 310 fewer, and put some of the savings into aviation readiness.[78] The plan has little transformational value for the Navy. Discomfort lies more with the Marines, who seem to have been strong-armed into cooperating. The Marines worry that the plan will force a smaller buy of the vertical takeoff (STOVL) F-35. They also worry about relying more on Navy pilots, who focus less than Marine Corps pilots on close air support – support that the Marines will need more of in a planned future where they move inland fast and thus bring less artillery. For the Navy, the plan involves slight operational change but nothing indigestible.

An additional organizational shift in the Navy is the Fleet Response Plan. The plan is a concession to the rise of surge (the idea of using the fleet in bulk for short periods of warfare rather than continual patrolling missions) as a force justifier, as opposed to presence. CNO Vern Clark released the plan in 2003, probably with prodding from OSD. Clark developed the plan to make more carriers deployable in less time. Carriers are now ready for redeployment within nine months from return instead of eighteen. This means that of those in port, four are ready to go back out immediately, two more within ninety days.[79] To accomplish the shift, more money must be spent on readiness, which means supplies like spare parts, the tractors that pull planes on deck, and munitions. Before the plan, the Navy had swapped these things onto the carriers as they went out, avoiding expensive redundancy.[80] The Fleet Response Plan is a considerable change, and a slight shift in acquisition, but not innovative. It did not fundamentally alter the role of carriers, nor threaten any community.

In October 2007, the Navy issued a new strategy, which seems likely to alter little about Navy force structure and planning.[81] The strategy's production was likely a consequence of the Navy's concern that current US defense priorities have a limited place for naval forces, despite the Navy's performance in the conventional phases of the wars in Iraq and Afghanistan. Counterinsurgency is manpower-intensive. Naval forces play only a small role. The Pentagon recently decided to increase the end strength and budget share of the Army and Marine Corps, slicing tiny bits from the Air Force and Navy budgets.

Driven by these concerns, Mullen launched a series of initiatives to heighten the Navy's role in counterinsurgency, or the perception thereof. He presided over the creation of the Navy Expeditionary Combat Command (NECC), established in January 2006 to oversee the formerly disparate naval functions with some claim to relevance to these wars. That includes Seabees, who have engineering expertise useful in post-conflict situations, riverine forces, some logistics capabilities, and several thousand sailors assigned to force protection in Iraq and Afghanistan. The SEALs, the most relevant part of the Navy to counter-terrorism, are not included.[82]

The cooperative strategy casts the Navy as something of a lead sheriff in a global constabulary force, looking for migrants, terrorists, drugs, and weapons. The idea is essentially to police the world's oceans using commercial shipping information and advanced sensors, achieving "global maritime awareness."[83] Part of this vision is the Proliferation Security Initiative (PSI), an intelligence-sharing activity organized by the United States and meant to stop the spread of chemical, biological, and nuclear weapons and missiles at sea. However, the PSI – leaving aside the fact that it is unclear if it has seized anything above what could have been seized in its absence – might not require much from navies. Most weapon seizures can be made in port. Moreover, there have not been any reports of terrorists being captured on high seas. Nor is it clear how constabulary missions justify destroyers and cruisers. The cooperative Navy concept also claims that naval alliances are essential to commerce protection. The problem with this idea is that few states have an interest in and ability to disrupt global shipping. It is also unclear why this mission requires large ships rather smaller, faster ships.

Within the Navy, this constabulary strategy is opposed by admirals who prefer a Navy justified around larger-scale possibilities like war with Iran, North Korea, and especially China. This concept can be a called a surge Navy, one that swiftly moves capabilities to hotspots during war. These are at least wars that require unique naval capabilities, unlike small wars. Along with concepts like capabilities-based planning, these scenarios can be combined to require significant force structure and substantial roles for all the major naval communities. Yet even the most ambitious of these navies, China's, has procurement patterns – no working aircraft carriers, little ASW capability, weak area air defense capabilities (no AWACS-type system) – suggestive of area defense rather than challenging rivals in blue water.

Neither of these visions is particularly transformational. The constabulary force idea logically leads to the conclusion that the Navy should purchase lots of frigates tied into various ISR platforms to police the world. The procurement plan is focused on large ships, however, and even LCS is not optimized for this sort of work. The cooperative Navy appears to be another way to sell existing capabilities, only with an emphasis on surface ships rather than carriers or submarines. Surface officers, including Mullen and Roughead, are atop the naval hierarchy today and are sometimes referred to as the Aegis mafia. The 313-ship goal may reflect this reality, giving priority to surface programs over submarines and aviation.

The surge navy concept appears to come from the aviation community. This strategy would presumably lead to a slight shift in budget allocation in favor of aviators, not transform the Navy. Like the constabulary view, this strategy would adopt the language of transformation while harnessing network language and technology to existing functions.

What changed?

For the Navy, the post-Cold War period has been a time of considerable administrative change and technological progress, not doctrinal upheaval. There is some evidence of progress toward transformation's stated goals. Jointness has increased. Somewhat small ships – LCS – are being built, but their size is really that of a frigate, which is not that small. Within the Navy, spending on research, development, testing, and evaluation has increased, but not faster than growth in the overall naval budget (that is, RDT&E's share is not growing). The same can be said for C4ISR – Command, Control, Communications, Computers, Intelligence, Surveillance and Reconnaissance.[84] Communications systems among sensors and shooters brought cooperative engagement capability and allowed naval aircraft to engage in time-sensitive targeting with precision. Destroying a given target is considerably cheaper than it once was, at least via carrier-based aviation, even though overall budgets remain high. As noted, several of these changes might simply be the result of the end of the Cold War rather than revolutionary technology.

Meanwhile, the Navy has changed its structure somewhat to focus on land powers. Each community now competes to defeat shore-based targets. Carrier-based airpower has greatly improved its destructive capability. The fleet response plan makes carriers' striking power ashore more sustainable, as do the myriad of improvements in naval aviation, including the integration of Marine and Navy airpower. Expeditionary Strike Groups offer a new force package for attacking enemy shores. These changes are a response to both the international environment and competition with rival services. From the purview of organizational politics, however, the changes are not innovative. They largely protect existing doctrine and platforms.

Three organizational changes in the post-Cold War Navy do stand out as partially or potentially innovative.[85] First, LCS and destroyer's role in mine warfare should allow the retirement of the Navy's two types of mine-warfare vessels and the Oliver Hazard Perry-class frigates. This is at best mild innovation, given that the mine-warfare community has always been peripheral, LCS is essentially a new frigate, and the surface community does not lose ships. And it not yet clear that there will be a full LCS buy and that all these other ships will be scrapped on schedule. Second, the tactical aviation rivalry between the Navy and Air Force might cause creative destruction in an age of low budget ceilings and less restrained interservice rivalry. Today, budgetary plenty and jointness prevent that outcome.

Third is the submarine community's scramble for relevance. A perceived threat to submarines' role and budget caused their operators to find new missions. Submarines heightened their intelligence gathering efforts and gained the ability to carry SEALs and fire cruise missiles, competing with surface ships and aviation. This response is potentially innovative, but it is not clear that their new missions have cost other communities. In any case, the cause is international politics, namely the end of the Cold War, not the revolution in military affairs.

Overall, the Navy has not changed much since the Cold War. Platforms are upgraded more than replaced. Communities have not much undermined each other's relevance with doctrinal innovation. With three near-equal communities, the power structure of the Navy basically holds. Table 4.1, taken from a Congressional Budget Office report, makes this point.[86] It shows how many of each type of naval ship the United States had in 1990 and in 2006, and the percentage of the total number of ships this amounts to. The total number has declined by roughly half, but the percentages have barely changed; none exceed a 3 percent difference.

Explaining naval continuity

There are several reasons, which would probably suffice alone, for the perseverance of this status quo in the Navy. The first explanation is that there was no technological revolution to be had. The theory of transformation likely overstated the potential impact of the technologies it heralded. Networking is nothing new to the Navy. Sensor technologies, information exchange and processing, and

Table 4.1 Naval platforms after the Cold War

	1990		*2006*	
	Number of ships	*Percentage of fleet*	*Number of ships*	*Percentage of fleet*
Aircraft carriers	15	3	12	4
Ballistic missile submarines	35	6	14	5
Attack submarines	97	17	57 (includes SSGN)	20
Surface combatants	213	37	102	35
Amphibious	66	11	35	12
Combat logistics	60	10	32	11
Mine warfare and fleet auxiliaries	88	15	33	12
Total	574	100	285	100

precision serve the Navy's existing doctrines rather than overthrowing them. Cooperative engagement capability and improved data sharing with aircraft for airstrikes are examples. RMA technologies made the existing Navy better at its work, an impressive but not revolutionary achievement.

We cannot, however, credit technology alone for the failure of the revolution. Predicting whether technologies are revolutionary is impossible in the absence of their marriage to international and domestic politics. For example, tanks, airpower, and radios did not make blitzkrieg. Germany did. They employed old technologies in service of a strategy. Technology is the servant of politics, not the reverse.[87] New technologies facilitate revolutions in military affairs, but strategic necessity or cunning statesmen must always make them.

For three reasons, the Navy will not transform itself. First, greatly upsetting the power-sharing arrangement among the three naval communities would cause a fight that naval leaders have little incentive to seek. Weak communities like mine warfare can be downgraded, but the big three remain. Second, sunk costs make change expensive. The cost of ships and the intellectual capital to operate and create them gives the continuity a large cost advantage over change. Third, current ways of doing business have earned their political keep. The Navy's mix of missions and platforms and the jobs they provide in shipyards have earned steady Congressional support. Innovation threatens that support.

These factors mean that, like most organizations, the Navy tends to adapt to new missions warily but embraces change that does not disturb its internal dynamics.[88] It follows that absent some outside intrusion, the Navy's communities will try to preserve their function. They will embrace technologies to improve the execution of their task, including those usually called transformational, and avoid revolutionary change.

The next question is why outside forces did not much disturb the Navy's preferences. The Clinton Administration was buffeted by a Republican Congress eager to denounce it for meddling with the military and had limited control of the military from the outset. The Administration also ran the Pentagon at a time when threats were low – denying would-be civilian innovators the alarm that helps overcome organizational desires.[89] The Bush Administration's pre-September 11 emphasis on transformation was limited by its vague goals, but it does seem to have thrust LCS on a reluctant Navy.

There was a heightened perception of threat after 2001 and some acceleration of change as a result. However, the existing Navy was well prepared to deal with the bombing campaigns in Afghanistan and Iraq, so the conventional conflicts required no organizational upheaval. As for unconventional war and counter-terrorism, naval forces are not very relevant, whatever the rhetoric, so there was little obvious case for civilian-led reform of the Navy on that front. Compare the Navy, for example, with the FBI, where fighting terrorism required considerable reorganization.

Another prominent theory of organizational innovation says that organizations can be frightened into innovation if their broader sense of organizational purpose is threatened.[90] Civilians provoke such competition with strategies that select among services.[91] In neither Administration was there an articulation of a strategy that favored a particular service, and thus no competition and little innovation. The Navy had little incentive to please its civilian masters by innovating, unless we count its awkward efforts in recent years to define itself as a crucial part of counterinsurgency campaigns.

What's next?

The rise of these wars and counter-terrorism as the centerpiece of national security policy threatens the Navy with irrelevance. This shift is symptomatic of a truly revolutionary trend, one ignored by transformation theory: the steep decline of competitive military procurement among large states. Many states today perceive less risk than their predecessors did from neighbors and invest less in militaries to defend themselves. Reasons for this change include the economic shift away from industrial economies and the rise of nationalism, which undermined the utility of military conquest, settled borders, nuclear weapons, and changing ideas about using force.[92] Interstate war, which drove the creation of navies, has been in decline for decades. China notwithstanding, rich states are not buying navies as they once did to compete for wealth and power. That is the revolution that could shake the Navy.

What is the Navy to do? There is no escaping this geopolitical shift by claiming a place in small wars. They are rarely fought on water and little affected by sea-based guns, missiles, and aviation. Nor do terrorists justify much naval spending. After all, they rarely congregate enough to invite cruise missiles, cannot much threaten sea lanes, and certainly do not possess weapons that require Aegis defense systems.

The best story that the Navy can tell is that even if competitive military procurement is declining, Americans must still project power to defend shipping and allies from state threats, real and potential. The Navy's claim should be that it exists not only to balance against rival navies but also to deter their formation. The Navy should bill itself as insurance, and not the sort that provokes enmity by living in foreign lands. In other words, the United States is a maritime power that lacks the skill to run foreign states and the patience and interest to pacify them from insurgent violence. What Americans can do well is threaten enemies with various sorts of violence from the sea. That role no longer requires eleven carrier battle groups or 300 ships, but it is the best case that the Navy has.

Notes

1 The author thanks the following people for sharing their knowledge about the US Navy: Paul Nagy, Peter Swartz, Wayne Hughes, Jeffrey Canfield, Hank Gaffney, Michael Parkyn, Dave Ditallo, and especially Owen Cote.

2 This is not a criticism of the Navy. Organizations dedicated to a purpose are rightfully hesitant to abolish proven methods of providing it. Evolution is a sensible way of dealing with changes in the domestic and international political environment. Organizations that continually revolutionized themselves would damage their effectiveness. Jeremy Shapiro, "Information and War: Is it a Revolution?" in *Strategic Appraisal: The Changing Role of Information in Warfare*, Zalmay M. Khalilzad and John P. White, eds. (Santa Monica, CA: RAND, 1999), p. 142.

3 Shapiro, "Information and War," p. 138. Shapiro notes that revolutions in military affairs are nearly impossible to predict. For an example of a work that mixes prediction and prescription, see William Owens with Ed Offley, *Lifting the Fog of War* (New York: Farrar, Straus, and Giroux, 2000), which at times appears to argue for the inevitability of these changes and at others to plead for their adoption. Like other authors, Owens also argues that the United States must embrace transformation because future adversaries like the Chinese certainly will, never explaining why it is difficult for us and inevitable for them.

4 Of these three realms, information technology and communication are the most important in network-centric warfare. Arthur Cebrowski especially links the advent of network-centric warfare to broader changes in modern society – to the idea that the industrial age has been replaced by the information age. James Blaker, "Arthur K. Cebrowski: A Retrospective," *Naval War College Review*, Vol. 59, No. 2 (Spring 2006), pp. 129–145.

5 Arthur K. Cebrowski and John J. Garstka, "Network-Centric Warfare: Its Origin and Future," *US Naval Institute Proceedings*, Vol. 124, No. 1 (January 1998), pp. 28–35.

6 John Arquilla, *Swarming and the Future of Warfare* (Santa Monica, CA: RAND, 2000).

7 Owens, *Lifting the Fog of War*, p. 23.

8 On the case for Streetfighter, see Arthur K. Cebrowski and Wayne P. Hughes, "Rebalancing the Fleet," *US Naval Institute Proceedings*, Vol. 125, No. 11 (November 1999), pp. 31–34; Wayne P. Hughes, Jr., "22 Questions for Streetfighter," *US Naval Institute Proceedings*, Vol. 125, No. 2 (February 2000), pp. 46–49. For an opposing view, see Richard B. Brawley, "Streetfighter Cannot Do the Job," *US Naval Institute Proceedings*, Vol. 128, No. 10 (October 2002), pp. 66–69. Each Streetfighter was envisioned to weigh about 500 tons and cost about $90 million.

9 Owens, *Lifting the Fog of War*, pp. 57–59.

10 Ibid., pp. 89–94. On Navy–Air Force cooperation, see John L. Barry and James Blaker, "After the Storm: The Growing Convergence of the Air Force and Navy," *Naval War College Review*, Vol. 54, No. 4 (Autumn 2001), pp. 117–133.

11 Andrew F. Krepinevich, Jr., "Military Experimentation: Time to Get Serious," *Naval War College Review*, Vol. 54, No. 1 (Winter 2001), pp. 76–89; Thomas G. Mahnken, "Transforming the US Armed Forces: Rhetoric or Reality?" *Navy War College Review*, Vol. 54. No. 3 (Summer 2001), pp. 85–89; Owens, *Lifting the Fog of War*, p. 209.

12 Owens, *Lifting the Fog of War*, pp. 225–226.

13 On innovation in militaries, see Harvey Sapolsky, *The Polaris System Development: Bureaucratic and Programmatic Success in Government* (Cambridge, MA: Harvard University Press, 1992); Barry R. Posen, *The Sources of Military Doctrine: Britain and Germany Between the World Wars* (Ithaca, NY: Cornell University Press, 1984); Stephen P. Rosen, *Winning the Next War: Innovation and the Modern Military* (Ithaca, NY: Cornell University Press, 1991); Owen R. Cote, Jr., *The Politics of Innovative Military Doctrine: The US Navy and Fleet Ballistic Missiles* (PhD Dissertation, Massachusetts Institute of Technology, 1996); Harvey Sapolsky, "On the Theory of Military Innovation," *Breakthroughs*, Vol. IX, No. 1 (Spring 2000), pp. 35–39.

14 Joseph Schumpeter, *Capitalism, Socialism, and Democracy* (New York, NY: Harpers, 1975).

15 Ballistic missile defense has restored some of surface's ships independence in recent years.

16 Cote, *The Politics of Innovative Military Doctrine*, pp. 8–9. Some writers use more inclusive definitions of innovations that include less disruptive changes, like embracing new technology. See, for example, Williamson Murray and Alan Millet, eds., *Military Innovation in the Interwar Period* (Cambridge, UK: Cambridge University Press, 1998).

17 The Department of the Navy includes the Marine Corps, but the Corps is highly independent, controlling its own personnel and budget. For this reason, the Marines Corps is addressed in its own chapter, as it would no doubt prefer. Smaller, weaker communities in the Navy include mine warfare and logistics.

18 A carrier based naval air wing includes three squadrons of FA-18s, another FA-18 squadron provided by the Marines, an electronic attack squadron of EA-6B prowlers, which are used for jamming enemy radar and communication, one squadron of E-2Cs, which provide early warning radar defense, a squadron of C-2 cargo aircraft, and one squadron each of antisubmarine and sea control helicopters. F-35 joint strike fighters are slated to replace FA-18s. Some naval airpower is land based, particularly during land wars. That is the case today with the war in Iraq.

19 D. Robert Worley, *Shaping US Military Forces: Revolution or Relevance in the Post-Cold War World* (London: Praeger, 2006), pp. 143–150.

20 Rosen, *Winning the Next War*; James. Q. Wilson, "Innovation in Organization: Notes Towards a Theory," in *Approaches to Organizational Design*, James D. Thompson, ed. (Pittsburgh, PA: University of Pittsburgh Press, 1971), pp. 195–218.

21 For more on this problem, see Eugene Gholz and Harvey M. Sapolsky, "Restructuring the US Defense Industry," *International Security*, Vol. 24, No. 3 (Winter 1999–2000), pp. 5–51. For a short explanation of the politics of shipyards, see John Shepard, Jr. and Harvey Sapolsky, "Five Shipyards Too Many," *Washington Post*, November 8, 2005.

22 Interview, Peter Swartz, Center for Naval Analysis, July 31, 2006. Defense budgets had already begun to rise at the end of the Carter Administration.

23 Ibid.

24 Ronald O'Rourke, "Navy Ship Procurement Rate and the Planned Size of the Navy: Background and Issue for Congress," Congressional Research Service (March 1, 2002), p. 2.

25 Aegis is an integrated system of radar and weapons controls used to track and destroy enemy targets at sea. The AN/SPY-1D Phased Array Radar system can simultaneously search for incoming aircraft and missiles, track them and target them with missiles. Destroyers and cruisers that employ the system are often referred to as Aegis Destroyers or Aegis Cruisers.

26 Soviets tactics were improving enough that the American anti-submarine warfare capability may have become ineffective had the Cold War not ended. Owen R. Cote, Jr., *The Third Battle: Innovation in the US Navy's Silent Cold War Struggle with Soviet Submarines* (Newport Paper #16, US Naval War College, 2003).

27 Lorna Jaffe, *The Development of the Base Force 1989–1992* (Joint History Office, Office of the Chairman of the Joint Chiefs of Staff, 1993). In private, Powell was fond of saying how there were no enemies left to justify the force structure (Swartz interview). Swartz was an aide to Powell. Powell was famously quoted: "I'm running out of demons. I'm running out of enemies. I'm down to Castro and Kim Il Sung." Quoted in "Cover Story, Communism's Collapse Poses a Challenge to America's Military," *US News and World Report*, October 14, 1991, p. 28.

28 William Kaufmann, *Assessing the Base Force: How Much Is Too Much?* (Washington, DC: Brookings Institution, 1992).

29 Swartz interview. Three straight Chiefs of Naval Operations were submariners: James Watkins, 1982–1986, Carlisle Trost, 1986–1990, and Frank Kelso, 1990–1994.

30 Ibid.

31 Ibid.

32 The air war kicked off just as the Pentagon cancelled the A-12 in January 1991, the Navy's stealth fighter-bomber, meant to replace the A-6.

33 Interview with Hank Gaffney, Center for Naval Analysis, July 31, 2006.

34 Andrew Krepinevich, Jr., *The Bottom-Up Review: An Assessment* (Washington, DC: Defense Budget Project, 1994).

35 Gaffney interview.

36 The ATO could be sent via satellite link to most Air Force units. Fleet UHF satellites on carriers lacked the necessary capacity. Norman Friedman, *Seapower and Space: From the Dawn of the Missile Age to Net-Centric Warfare* (Annapolis, MD: Naval Institute Press, 200), p. 284.

37 There was a fifth shortcoming. The Navy's F-14 and F-18 fighters lacked redundant target recognition systems. The Air Force's F-15s had two Identification, Friend or Foe (IFF) systems, which send an electronic coded signal to other aircraft. The rules of engagement for the Gulf War, set by the Air Force JFACC, stipulated that two IFF efforts had to be made prior to engaging (attacking) enemy aircraft. This rule prevented F-14s from getting kills. Many naval aviators saw the rules as an effort by the Air Force to thwart naval aviation and promote the Air Force's capabilities. P. Mason Carpenter, *Joint Operations in the Gulf War: An Allison Analysis* (Maxwell-Gunter AFB, AL: Air University, Thesis, 1995), pp. 11–20.

38 Friedman, *Seapower and Space*, p. 284.

39 Swartz Interview. The Tailhook sexual harassment scandal also hurt morale during this period. Tailhook is an annual naval aviators' party and meeting, and therefore principally hurt the carrier aviation community. The notorious meeting in held 1991 focused on assessing the Gulf War.

40 This is slightly disingenuous, since Naval aviation usually requires Air Force (or British) tanking to refuel, and carriers are tethered to supply provided by the Air Force (Gaffney interview). On the other hand, several countries, notably in the Middle East, have proved far more willing to allow tankers and reconnaissance to take off and land in their territory than fighters and bombers. Owen R. Cote, Jr. *The Future of Naval Aviation* (Cambridge, MA: MIT Security Studies Program, 2006), pp. 14–16.

41 Swartz interview.

42 For an overview see, John Hattendorf, ed., *US Naval Strategy in the 1990s: Selected Documents* (Newport, RI: Naval War College Press, 2006).

43 Cote, *The Future of Naval Aviation*, pp. 50–51.

44 Receiving the ATO required changes to naval antenna used to receive satellite signals. The diameter of the antenna was increased, providing higher gain. Friedman, *Seapower and Space*, p. 285.

45 Benjamin Lambeth, *American Carrier Air Power at the Dawn of a New Century* (Santa Monica, CA: RAND, 2005), p. 3.
46 Ibid., p. 4.
47 Tomahawks saw plentiful use in the decade (Swartz interview).
48 For a discussion of the arsenal ship's proposed capabilities, see Robert S. Leonard, Jeffrey A. Drezner, and Geoffrey Somner, *The Arsenal Ship: Acquisition Process Experience* (Santa Monica, CA: RAND, 1999), pp. 97–105.
49 Ibid., pp. 88–89.
50 Ibid., pp. 85–99.
51 Mahnken, "Transforming the US Armed Forces."
52 "Expeditionary Strike Group," Globalsecurity.org. Online, available at: www.globalsecurity.org/military/agency/navy/esg.htm.
53 "ASDS Mini-Sub Program Taking On Water," *Defense Industry Daily*, August 2, 2006. Online, available at: www.defenseindustrydaily.com/asds-minisub-program-taking-on-water-updated-01580/.
54 Submarines have traditionally collected information on enemy submarines using sonar, but have also used periscopes for photographic intelligence, and tapped into undersea cables to intercept communications, among other missions. Jeffrey Richelson, *The US Intelligence Community* (Cambridge, MA: Ballinger, 2nd edn, 1989), p. 191. On the increased role of intelligence gathering as an attack submarine mission, see Rick Barnard, "Congress Should Reassess Attack Boat Program as Mission Days Dwindle," *Sea Power*, April 20, 2003; Vice Admiral E. P. Giambastiani, Commander Submarine Force US Atlantic Fleet Commander Submarines Allied Command Atlantic, "Statement Before The Senate Armed Services Committee Seapower Subcommittee," April 13, 1999. Online, available at: www.fas.org/man/congress/1999/giambast.htm.
55 In the long term, submarines' prospects are bolstered by the argument that the proliferation of missile technology and surveillance satellites will so increase the vulnerability of surface combatants that navies will be driven underwater. Friedman, *Seapower and Space*, p. 308.
56 Eric Labs, "Options for the Navy's Future Fleet," Congressional Budget Office Report, May 2006, p. xvi.
57 There are also proposals to convert some Trident ballistic missiles on SSBNs to a conventional role – the so-called tactical Trident. Advocates claim that ballistic missiles' added range and speed makes them useful in striking fleeting targets. The conversion would be controversial. The worry is that states like Russia and China might observe the launch and interpret it as a nuclear attack. Elaine M. Grossman, "Pentagon Eyes Bunker-Busting Conventional Ballistic Role for Subs," *Inside the Pentagon*, June 27, 2002.
58 Les Aspin, Secretary of Defense, *Report on the Bottom-Up Review* (US Department of Defense, 1993). By 1994, the Navy realized it could only fund 330 ships. "An Assessment of the Bottom Up Review," *Hearing Before the Military Forces and Personnel Subcommittee of the Committee on Armed Services*, US House of Representatives (Washington, DC: Government Printing Office, 1994).
59 Krepinevich, "The Bottom-Up Review."
60 Edward Rhodes, "'...From the Sea' and Back Again: Naval Power in the Second American Century," *Naval War College Review*, Vol. 52. No. 3 (Spring 1999), pp. 13–54; Daniel Goure, "The Tyranny of Forward Presence," *Naval War College Review*, Vol. 54, No. 3 (Summer 2001), pp. 11–24.
61 Eric Labs, "Building the Future Fleet: Show Us the Analysis!" *Naval War College Review*, Vol. 57, No. 3–4 (Summer/Autumn 2004), pp. 138–146.
62 Sam J. Tangredi, "The Rise and Fall of Naval Forward Presence," *US Naval Institute Proceedings*, Vol. 125, No. 5 (May 2000), pp. 28–33.
63 Friedman, *Seapower and Space*, pp. 298–299.
64 Vern Clark, "Sea Power 21: Projecting Decisive Joint Capabilities," *US Naval Institute Proceedings*, Vol. 128, No. 10 (October 2002), pp. 32–41.

65 Friedman, *Seapower and Space*, p. 288.
66 Lambeth, *American Carrier Air Power*, p. 88.
67 The retirement of the A-6 in 1997 and the cancellation of its planned replacement, the stealthy A-12, after the Cold War, left the Navy with less striking range than it had in the 1980s. The efforts undertaken in the early 1990s could not quickly overcome this problem. For a good discussion of the Navy's efforts to improve the range of carrier-air wings, see Thomas Ehrhard and Robert O. Work "The Unmanned Combat Air System Carrier Demonstration Program: A New Dawn for Naval Aviation?" Center for Strategic and Budgetary Assessments, Backgrounder, May 10, 2007.
68 Ibid., p. 23.
69 Lambeth, *American Carrier Air Power*, p. 11.
70 Ibid., p. 52.
71 According to a study produced by the Navy in 2001, a 2001 carrier air wing could strike 683 aimpoints (or targets) per day at a range of 200 nautical miles, whereas in 1989 a carrier air wing could destroy only 162 aimpoints at the same range. A carrier air wing will be able to destroy 1,080 targets a day at the same range by 2010, according to the study. These figures are based on a series of assumptions and are useful mainly for comparison. As cited by Ehrhard and Work, "The Unmanned Air System Carrier Demonstration Program," pp. 22–23.
72 Christopher Bolkcom, "F-35 Joint Strike Fighter (JSF) Program: Background, Status, and Issues," Congressional Research Service (June 2006), p. 8.
73 See for example, Duncan Long and Stuart Johnson "The Littoral Combat Ship: From Concept to Program," Center for Technology and National Security Policy, National Defense University, Case Studies in National Security Transformation, Number 7, March 2007, p. 8.
74 Steve Liewer, "Costs Snag New Naval Combat Ship," *San Diego Union Tribune*, April 23, 2007. Costs should drop if production accelerates and economies of scale are achieved, but even then the ships will be far more expensive than planned.
75 The Perrys will exit the fleet on the heels of Spruance class destroyers, which are being replaced by Arleigh Burkes. Labs, "Options for the Navy's Future Fleet," p. xvi.
76 Ronald O'Rourke, "Naval Littoral Combat Ship (LCS): Background and Issues for Congress," Congressional Research Service (August 2006), p. 1.
77 Eric Labs estimates that to reach its goal of 313 ships by 2035, the Navy's annual shipbuilding budget will have be to be $21 billion, 70 percent more than the average between 2000 and 2005 in constant 2007 dollars. Labs, "Options for the Navy's Future Fleet," p. 21.
78 Christopher Bolkcom and Ronald O'Rourke, "Navy-Marine Corps Tactical Air Integration Plan: Background and Issues for Congress," Congressional Research Service (April 2003). The Marines have twenty-one tactical aircraft squadrons and the Navy thirty-nine. The difference between what is bought and what is operated stems from the fact that some are usually in overhaul, and some will likely be lost to accidents and maybe combat.
79 Lambeth, *American Carrier Air Power*, pp. 59–64.
80 A problem with the Fleet Response Plan is that heightened readiness costs other accounts, like shipbuilding. Its wisdom depends on how often surge actually occurs and how urgent the need really is. There is a school of thought that says since the fleet only surges about once a decade, you might as well adjust on the fly rather than pay a lot for enough equipment to keep every carrier's readiness high (Gaffney interview).
81 "A Cooperative Strategy for 21st Century Seapower" (October 2007). Online, available at: www.navy.mil/maritime/MaritimeStrategy.pdf.
82 SEALs are more a part of the Special Forces community than the Navy. See Chapter 7.
83 John G. Morgan, Jr., and Charles W. Martoglio, "The 1,000-Ship Navy: Global Maritime Network," *US Naval Institute Proceedings*, Vol. 132, No. 11 (November 2005), pp. 14–17.

84 "Long-Term Implications of Current Defense Spending: Summary Update for Fiscal Year 2007," Congressional Budget Office (October 2006), p. 17. This result is a mixed bag in evaluating whether transformation has occurred. Its advocates call for more spending on these budget items, but the increase is less meaningful when the share of the whole is constant. In the military as a whole, research and development spending has held steady at about 13 percent of the defense budget during the Bush Administration. "Defense Budget by Title," Center for Strategic and Budgetary Assessments. Online, available at: www.csbaonline.org/2006–1/2.DefenseBudget/By_Title.shtml. See also, Steve Kosiak, *Analysis of the FY 2008 Defense Budget Request*, Center for Strategic and Budgetary Assessments, pp. 19–20. Online, available at: www.csbaonline.org/4Publications/PubLibrary/R.20070607.Analysis_of_the_FY/R.20070607.Analysis_of_the_FY.pdf.

85 Another potentially transformative change is the advent of a ballistic missile defense mission for surface ships. The Navy is outfitting its Ticonderoga-class cruisers and Arleigh Burke-class destroyers to track and potentially intercept ballistic missiles using upgraded AN/SPY-1 radar and Standard missiles. This role would be innovative if it took or limited the mission of another element of the nation's missile defense infrastructure, such as ground-based interceptors. There is no sign of that occurring.

86 Labs, "Options for the Navy's Future Fleet," p. 2.

87 Keir Lieber, *War and the Engineers: The Primacy of Politics over Technology* (Ithaca, NY: Cornell University Press, 2005).

88 James Q. Wilson, "Innovation in Organization."

89 Posen, *Sources of Military Doctrine.*

90 Sapolsky, "On the Theory of Military Innovation."

91 Cote, *The Politics of Innovative Military Doctrine*, pp. 70–94.

92 On this point, see Sean M. Lynn-Jones and Steven E. Miller, eds., *The Cold War and After: Prospects for Peace* (Cambridge, MA: MIT Press, 1994), particularly John Mueller, "The Essential Irrelevance of Nuclear Weapons: Stability in the Postwar World," pp. 45–69; Carl Kaysen, "Is War Obsolete? A Review Essay," pp. 81–103; and Stephen Van Evera, "Primed for Peace: Europe After the Cold War," pp. 193–243.

5 Evolution in the post-Cold War Air Force

Technology, doctrine, and bureaucratic politics

Sanford L. Weiner

The Air Force had an easier time in the 1990s than the other services. The astounding success of Desert Storm brought greatly increased respect for airpower as a component in warfare. In addition, the technologies thought to be central to a "Revolution in Military Affairs" (RMA) – stealth, precision munitions, more effective communications – were all close to the service's existing core concerns. So it was much easier to shape the utilization of these technologies to existing Air Force organizational interests, producing evolution rather than transformation.

However, Desert Storm was supposed to be only the beginning of the new era of airpower dominance. It has not worked out that way. Each combat experience since 1991 has brought more constraints and complications than Desert Storm did, and the Air Force today is as perplexed by counterinsurgency as the rest of the American military.

Throughout this period, the Air Force has sought to continue the massive modernization campaign of all its major platforms that began during the Reagan era. Many of these choices were only distantly related to national strategic needs, but the budget process has been the only constraint. This tension will continue as the service seeks tens of billions dollars for the F-22 and F-35 fighters, which now look to many like the continuation of Cold War thinking rather than effective transformation.

From autonomy to precision

The Air Force began with the rhetoric of fundamental change. From World War I onward airpower enthusiasts saw a radical technology which, if properly developed and applied, would transform modern warfare. But General Pershing and other Army leaders emphatically rejected that vision and saw airplanes as marginal adjuncts to ground forces: "He believed that the Air Force's main mission should be 'to drive off hostile airplanes and procure for the infantry and artillery information concerning the enemy's movements'."[1] In response to this opposition, the airpower theorists developed their autonomy triad.[2] Only those who appreciated the full potential of airpower should be in charge of developing and managing it, they argued. Thus a separate, autonomous organization was required. Second, air resources would be most useful in combat as a centralized autonomous strike force, not frittered away in support of multiple ground force units. Finally, to justify this independence, a doctrine was needed that prescribed a unique mission that only airpower could achieve: strategic bombing deep behind enemy lines. It's important to note that this doctrine was as much theology as strategy. It would be fifteen to twenty years before bombers even began to gain the capabilities needed. But the advocates chose to focus on the long-term potential of an advancing technology.

The Army Air Corps tactical school spent the 1930s elaborating this doctrine. By the eve of World War II, their strategy document outlined how "precision strategic bombing" would be the decisive factor in the next war.[3] The latest in precision technology, the Norden bombsight, would allow bombing of key logistics nodes. This effort would then cripple the enemy and force an early surrender, or make a ground campaign relatively easy – the attrition trench warfare of World War I need not be repeated.

Of course the actual Air Force impact on the war was more mixed and controversial.[4] Although it had a major impact in many combined arms engagements, it was only late in 1944, when German oil supplies were severely curtailed, that bombing began to have a strategic impact. By then, numerous other forces were bringing the war to a close.

Instead of revising their theories, strategic bombing advocates merely advanced them into the future, when the technology would match their vision. One of the clear issues during World War II was that the bombing was "precise" in name only (see below, Table 5.1). Thus, even before the advent of smart weapons, the Air Force welcomed changes in technology and training that increased bombing accuracy. Richard Hallion has calculated that accuracy increased significantly, first in Korea and then again in Vietnam.[5] Table 5.1 shows the number of 2,000-pound bombs an aircraft needed to drop from medium altitude to have a 90 percent probability of hitting a 60×100 foot target in World War II, Korea, and Vietnam.

The first laser-guided bombs (LGBs) arrived in combat in Vietnam in 1972. In a famous demonstration, LGBs destroyed in one afternoon the Thanh Hoa bridge in North Vietnam, which had withstood seven years of attacks with unguided munitions at the cost of dozens of aircraft and killed or captured crew. Since that time, the Air Force has generally welcomed further PGMs – at least

Table 5.1 Historical bombing accuracy

	Bomb	*Planes*	*CEP (feet)*
World War II	9,070	3,024	3,300
Korea	1,100	550	1,000
Vietnam	176	44	400

Note
(CEP = Circle of Error Probability; 50 percent of bombs will hit within the circle)

those meant for existing platforms. They materially advance a core interest of the service.

In his foreword to the Air Force official history, Hallion observes that "more than any other military organization, the US Air Force searches out and listens to the experience history offers."[6] This contention needs to be divided into two categories. The Air Force is indeed very conscious of its organizational history, especially its long-time pursuit of respect and autonomy. But for actual combat history the Air Force has been very skilled at framing its own interpretation of events as the key policy lessons. These lessons are often a very partial reflection of the actual events. Thus from 1991 to the beginning of 2003, each major conflict was seen as an opportunity to demonstrate the growing precision and potency of airpower, and the qualifications were downplayed.

The following sections describe the debates over the lessons from each of the major conflicts from the Gulf War on, and the key platforms and other technology employed each time.

The Gulf War

"We are not preparing the battlefield, we are destroying it" (Lt. Col. (now Gen.) David Deptula).[7] During the 1960s and 1970s the Strategic Air Command concerned itself with the very different strategies posed by the deployment of nuclear weapons. The Tactical Air Command (TAC) did most of the Air Force bombing in Vietnam. They developed new bombing approaches that highlighted incremental battlefield interdiction. These were still the dominant ideas in the 1980s when Colonel John Warden and other reformers began to argue that precision weapons now made decisive strategic bombing possible at last.[8] In both personality and doctrine, Warden was a controversial figure before the Gulf War ("an iconoclast to his critics, a visionary to admirers").[9] The head of TAC disliked him, but he had other strong supporters among other senior leaders.

In the fall of 1990 he developed his "Instant Thunder" plan for the upcoming Gulf War. Warden made one trip to Saudi Arabia to present his plan, and was soon ordered to leave the country. But David Deptula, his deputy, was retained at CENTCOM and became a key planner for the actual air campaign.[10]

As it turned out, the Gulf War provided the strongest evidence to date of the effectiveness of airpower. Gordon and Trainor reflect the general consensus:

> The Air Force did deliver on its promise to make any ground offensive a walkover … the air attacks made it impossible for the Iraqis to mount an effective defense. Airpower crippled the Iraqi war machine. It neutralized sophisticated air defense systems, destroyed bridges and road junctions, destroyed the Iraqi artillery, and made it difficult for Iraq to maneuver forces on the battlefield.[11]

As the weeks went by, Iraqi divisions were decimated, tank by tank, until many had less than 50 percent of their initial fighting strength, and little offensive capability.[12]

Indeed, the Air Force view is that the Army did little more than accept the surrender after the Air Force had won the war. They discounted critics who pointed out that the war's circumstances were uniquely suited to airpower, and might not occur again (combat in open desert, with the enemy using Soviet-style World War II tank formations, without any serious air defenses).

Precision weapons amounted to only about 10 percent of the total munitions used, but had a disproportionate impact on targets successfully destroyed. They were credited with over 60 percent of the total damage. This battlefield success became the rationale to move forward with adding stealth and precision to the entire Air Force. (The laser-guided bombs were being carried on aging F-111s.)

Colonel Warden and other airpower theorists then extrapolated from the Gulf War to a future conflict where all the US weapons would be precise. In their vision, the original promise of strategic bombing, so often anticipated and so long delayed, would now be possible. Dramatic and simultaneous attacks on all the enemy's key military and economic centers would coerce an early surrender; or cut off the regime leadership from control of its forces, which would lead to the same result. Warden was unswayed by academic critics like Robert Pape who argued that the most effective use of PGMs would be on enemy forces in the field, not the leadership in the capital.[13] With victory, Warden's ideas are now official Air Force doctrine (see below, pp. 108–109).[14]

The Warden/Deptula initiatives bear some resemblance to the innovation scenarios described by Steve Rosen (see Chapter 1, p. 7). As he describes, this was a group of middle ranking officers seeking support from particular superiors for their ideas. But the critical difference is that the "reformers" wanted to revive and expand traditional Air Force doctrine, which was much easier than pushing for truly innovative strategies.

Modernization and new technologies

By 1991, the Soviet Union, the superpower whose forces our weapons were designed to counter, had collapsed. The forces of Iraq, a major regional power, had been smashed. Post-Cold War budgets would be tighter. Yet the Air Force argued that most of the weapons that had just been used to gain the victory were insufficient and needed replacement with superior ones.

Bombers

The new F-117 had flown straight through Baghdad, unimpeded by any air defense measures, thus proving that stealth technology worked as intended in a battlefield situation. One policy conclusion from this success could have been that, in the future, all of the aircraft leading an initial attack should also be stealthy, although those that were utilized later, after the air defenses were neutralized (according to Air Force doctrine), would not need that very expensive capability. The Air Force chose the opposite position – it wanted as many different aircraft as possible to be made stealthy.

There had been no new heavy bombers from the time McNamara cancelled the B-70 in 1961 until the 1980s. The B-1 had been designed in the Nixon administration, cancelled by Carter in 1977, and then revived by the Reagan Pentagon in 1981. At that point the Air Force chose to go ahead with the soon-to-be outdated B-1 while it awaited progress on stealth technology, but Congress imposed a stringent program cap of about $20 billion. One hundred new aircraft were delivered in 1986–1988. But built under intense time and cost pressures, the defensive avionics on the planes did not work, and took years to be fixed.[15]

While this fiasco was taking place, development work on the completely new, stealthy B-2 continued and, after the Gulf War, the impetus was even stronger. The Air Force had originally conceived of a fleet of another 100 plus planes, but as costs rose to $2 billion a copy, Congress finally limited the total buy to twenty-one planes – at a cost of over $40 billion. Both the B-1 and B-2 had been originally designed as nuclear bombers. But in the mid-1990s, pressed by Congress, the Air Force gave them a conventional bomb load capability as well.

Fighters

Air Force officers had been very unhappy with the available fighter technology during the Vietnam War. The F-105 had been originally designed as a nuclear bomber, and the F-4 was a Navy-designed fleet defense interceptor.[16] As the fighter side of the Air Force grew more powerful after Vietnam, they pushed hard for two new fighter designs.[17] The F-15, which entered service in 1976, was a pure air superiority fighter. The F-16, which followed in 1980, could be used for both air combat and strike functions.[18] Throughout the 1980s, each design had been enhanced and upgraded. (Even the F-15E now has a strike capability.)

With the new technology available in the 1990s, the Air Force wanted to keep the same functionality in each plane, but make them stealthy as well. The stealthy air superiority fighter became the F-22. Its original specifications date from before the fall of the Soviet Union, so it has been designed to defeat enemy fighters that do not exist. The stealthy multi-mode design became the F-35. Because it has been designed to suit the needs of the Navy and Marines as well, the F-35 will be a "Joint Strike Fighter."

The Air Force, working with DARPA, also tried more exotic stealth combinations during the 1980s and 1990s. While JSTARS was being developed as an air-to-ground radar plane, there was a parallel black project to develop a stealthy radar plane.[19] After considerable amounts of money were spent, the project was abandoned as expensive and unwieldy.

JDAM

The Air Force leadership also gave careful consideration to how to continue to expand and upgrade the precision technologies. Cruise missiles had proven very effective, but at a cost of $1 million or more each, they were reserved for the highest-value or most difficult targets. Laser-guided bombs had also proved their utility, but are ineffective in bad weather (and cost about $75,000 each). They also require the pilot or other laser designator to keep illuminating the target until impact, which can be complicated and sometimes dangerous.

However, the successful deployment of the Global Positioning System (GPS) during the late 1980s offered a new technological opportunity.[20] The Air Force began a research project to attach GPS receivers to conventional bombs (JDAMs). With navigation information from the GPS, dumb bombs could gain active guidance and near-PGM accuracy, under any weather conditions. The design requirement called for relatively inexpensive (under $40,000 each) strap-on kits that could be used with the large inventory of existing bombs.[21]

Looking back, JDAMs are now thought of as one of the central elements of the movement for precision, because they are carried on numerous fighters and bombers. However, for much of their development they were not seen that way at all. The Air Force was thinking about hundreds of thousands of units, worth billions of dollars, at a time when the baseline defense budgets were being cut. In a very uncharacteristic policy, low cost was made a key design parameter. The original design proposals were more than $60,000 per unit, far beyond the intended budget.[22]

The JDAM project was made one of the demonstration projects for the new Clinton administration's plans for the reform of defense procurement. Most of the design specification and military specification part requirements were waived. The contractor would have substantial freedom to achieve the final performance requirements as it saw fit. But there was strong pressure to find commercial, off-the-shelf components to bring the cost down. The prime contractor and the principal subcontractors all spent considerable effort reviewing the components one by one. The key computer chip, for example, came from the same Motorola family then used for Apple computers. This effort succeeded, and the projected unit cost dropped by more than half.[23]

However, many looked at these efforts with a skeptical eye. GPS was adopted for use in critical military applications beyond navigation with great caution because it appeared fairly easy for an enemy to jam it. Backup systems were usually required. With a projected circular error probable of 13 meters, JDAMs would be somewhat less accurate than the systems they hoped to supplant. (Further upgrades have now reduced that figure by more than half, but they came later.)[24]

In addition, military engineers were uncomfortable with the emphasis on cheap, commercial components. They felt that combat situations imposed stresses on materials far beyond those experienced in civilian activities. Their concerns turned out to be partially justified. Subject to the intense forces on a bomber's external pylon while the plane performed a high-speed, low-altitude maneuver, JDAM parts vibrated and even cracked. The General Accounting Office highlighted the deficiencies, and the Pentagon's own testing officials required more re-engineering before giving the approval for general procurement. The JDAM Project Office had to upgrade some of the key components and accept less dramatic (but still substantial) cost savings.[25]

Kosovo

The initial reaction to Operation Allied Force in Kosovo and Serbia in 1999 was again a celebration of the potency of airpower. After seventy-three days of bombing, Serbian president Milosevic capitulated, without a ground soldier in sight. Within a year, he would be out of power and heading for a war crimes tribunal. The NATO Alliance, in its first joint combat experience, had a victory. John Keegan hailed this outcome:

> There are certain dates in the history of warfare that mark real turning points.... Now there is a new turning point to fix on the calendar: June 3, 1999, when the capitulation of President Milosevic proved that a war can be won by airpower alone.... The air forces have won a triumph, and are entitled to every plaudit they will receive and can look forward to enjoying a transformed status in the strategic community.[26]

Richard Hallion, in turn, cited Keegan with approval, and commented that Kosovo confirmed the new role of airpower heralded by Desert Storm.[27] The reactions to the newly available technology were also highly positive. The B-2 bomber, retrofitted so it could carry up to sixteen JDAM bombs per sortie, had been able to attack sixteen different targets with reliable high accuracy. Though the B-2s flew only forty-five sorties, they accounted for a notable number of targets destroyed.

These sorties had all been non-stop thirty-hour missions from their home base in Missouri. Aided by numerous in-flight refuelings, the planes had never touched down until they returned to base. The Air Force had proclaimed that its mission was "Global Reach, Global Power," and it was demonstrating that it now had an expeditionary force that could reach any spot on the planet.[28]

However, as the details of the overall operation became known, a very different picture emerged.[29] Even Benjamin Lambeth of RAND, a long-time airpower enthusiast, concluded that NATO won despite its strategy, not because of it.[30] In 1995, so it was thought, a few days of bombing had persuaded Milosevic to sign the Dayton Accords, bringing a peacekeeping force to Bosnia. As the Serbs stepped up their attacks in Kosovo, American and European diplomats thought another brief bombing campaign would again change Serbian policy.

The initial three days of bombing, however, brought only defiance from the Serbs this time. Thus NATO slid into an open-ended bombing campaign with neither political nor military preparations. All of the NATO political leaders had a veto on all the potential targets, so few were approved initially. In the first month, many of the same targets were hit again, because the target list had been exhausted.

The decision-makers assumed that in the precision era, detailed political sensitivities should be honored, down to the level of individual aim points. At their request, 2,000-pound bombs were sometimes replaced with 500-pound ones to reduce the collateral damage. (The Dutch vetoed any attacks on the Presidential palace, because it contained a Rembrandt.) All this produced an anemic bombing effort. Lambeth calculated that Desert Storm dropped more bombs in its first twelve hours than Allied Force did in its first twelve days.[31]

When the political concerns were finally satisfied, the military planning also proved inadequate. The Combined Air Operations Center (CAOC), which was tasked with managing the air campaign, had never been properly staffed to do target evaluation and selection. Although new staff was hurriedly added, the targets were picked individually, on an ad hoc basis. There were none of the systematic target groups that had been worked out for Desert Storm and were supposedly the norm for an effective air campaign.[32]

In addition, there was a crucial policy disconnect in the chain of command. The NATO commander, General Clark, thought his mission was to stop the ethnic cleansing in Kosovo. He wanted the majority of the sorties devoted to that task. But his air commander, General Michael Short, was a strong partisan of strategic bombing. He repeatedly expressed the view that attacking Serb forces in Kosovo was unlikely to make much difference, and only attacking Serbia itself was a productive use of his forces. Since Clark had little faith in his subordinate's approach, they wound up debating even individual targets in their video conferences, as their subordinates looked on.[33]

The Serbs had a dispersed, mobile air defense system that only turned its radars on for twenty seconds at a time. This made it very difficult to target those radars. In the end, the Serbs shot down only two planes, but the Coalition was never able to remove the air defense threat. Thus nearly all bombing was done from above 15,000 feet. Since the Serb forces in Kosovo were organized in very small units and took full advantage of the heavy forests and mountainous terrain, the bombing had no significant impact on their ethnic cleansing.[34] (Even JSTARS, which had been so impressive displaying Iraqi tank formations in the desert in 1991, had much more trouble imaging single trucks on mountain roads.)[35]

With the Kosovo bombing clearly ineffective, General Short was finally given permission to bomb regime targets in Serbia itself in late April.[36] Meanwhile, NATO leaders drew ever closer to sending ground troops. Finally, on June 3, Serbia gave in. The cause of the capitulation remains unclear. For General Clark, it was the threat of ground forces that mattered most.[37] But General Short had no doubt about which coercion had the most impact. He was quoted as saying that if he had been given the go-ahead to "go downtown," the bombing alone would have ended the conflict a month earlier.[38]

Analysts remain divided over the decisive element in changing Milosevic's mind. Some believe it was the tangible damage caused by the bombing of Serbia. Others think it was the persistence of the overall bombing campaign, which conveyed the message that Allied resolve was strong. Still others agree with General Clark that only the threat of a ground campaign finally convinced the Serbs that they had no chance.[39]

But for the Air Force, the takeaway lesson from Kosovo was straightforward: that strategic bombing matched its potential, and that JDAM-equipped

B2s would bring a worldwide capability to wage such a campaign. Deployments of both bombers and bombs were further stimulated. But never again, they complained, should there be so many political constraints standing in their way.[40]

Afghanistan

The first reaction to the need for a bombing plan for Afghanistan was that, after years of civil conflict, there were extremely few targets left for a strategic bombing campaign to focus on. However growing precision weapon capabilities have also increased the capacity for effective close air support, and this became the hallmark of the Afghan fighting. (There were 17,000 munitions dropped, but only 120 fixed targets.)[41]

Using new real-time targeting communications loops and laptop computers, Air Force or even CIA forces on the ground could serve as forward air controllers. They provided laser targeting or GPS coordinates to the aircraft pilots, and the bombs decimated Taliban positions. The Northern Alliance forces proceeded to an unexpectedly easy victory. It was also one where airpower plus local forces made large US ground force deployments seem unnecessary.

But this was Rumsfeld's preferred war, not that of the Air Force. No strategic bombing and no coercion of enemy leaders was involved. Furthermore the airpower needed to be combined with some ground forces to have its full effect. In the later stages of the conflict (Tora Bora) air power could not make up for the weaknesses of local allies.

In addition, the Air Force for the first time had to share the victory with the Navy. Carrier-based forces, combined with Air Force tankers, had been on the scene before some of the Air Force strike planes, which were still waiting for landing rights. The Navy planes now carried the laser-guided bombs that they had been notably lacking in Desert Storm, and were beginning to deploy JDAMs as well. Indeed 70 percent of the combat sorties and 50 percent of the precision munitions had been carrier-based.[42] (Heavy bombers carry more bombs per sortie though, and the Air Force also dropped a considerable number of dumb bombs.) As late as 1999, Air Force partisans were still writing articles on "the Myth of the Carrier."[43] That storyline was now out of date.[44]

Iraq war

By 2003, the growing PGM inventories made Air Force planners increasingly confident about the impact of the next strategic bombing campaign. With the ability to strike precisely and simultaneously at a large target set, they foresaw enemy leaders coerced by attacks on national targets, or at least constrained by the decimation of their command and control system. They looked forward to a true "shock and awe" campaign.

The official Air Force Doctrine, revised in 2003, used almost the same words as in 1941:

> Early airpower advocates argued that airpower could be decisive and could achieve strategic effects.... While this view of airpower was not proved during their lifetimes, the more recent history of air and space power application, especially since the 1991 Persian Gulf War, has proven that air and space power can be a dominant and frequently the decisive element of combat in modern warfare. Air and space power is a maneuver element in its own right, co-equal with land and maritime power; as such, it is no longer merely a supporting force to surface combat.
>
> The prompt, continued, aggressive application of air and space power in the opening phase may actually constitute the conflict's decisive phase. Thus, this first phase need not be a precursor to a buildup of ground forces and conventional counterattack.[45]

But everything about this war, even in its major combat stage, was less pleasing to the Air Force than in 1991. The Joint Air Commander asked for a two-week air campaign to win it all, or at least soften up the Republican Guard units, before the Coalition ground forces advanced. However, this time the Army felt it could handle the Iraqi divisions on its own. Although they had happily waited for bombers to decimate tanks in 1991, now they suggested no advance air campaign at all. Indeed, they argued that a simultaneous air and ground assault would preserve the most tactical surprise.[46]

Meanwhile both Turkey and Saudi Arabia had refused permission for offensive strikes from their bases. This complicated the route planning, and highlighted the newly recognized utility of the offshore carriers.

As the war plan evolved, the initial air campaign dropped from two weeks to one week to a few days. In the end, two misleading pieces of intelligence led to the ground war starting prematurely. Because the CIA reported that they knew where Saddam was staying, two F-117 sorties and dozens of cruise missiles bombed the residence ahead of the intended plan. And the ground troops crossed into Iraq early because of reports that the oil fields were about to be sabotaged. (Neither intelligence report was accurate.)[47]

Because the full-scale air attack was too complicated to move up, it didn't start until day three of the war. The target list was considerably restricted compared to 1991, though. This time the plan was to preserve the main bridges for the use of the Coalition forces advancing on Baghdad. Similarly, most of the electrical network was also being spared, to assist in the occupation/reconstruction to come.[48]

So there was neither coercion at the strategic level nor capitulation. Airwar theorists have written much about influencing the psychology of the enemy, but they had little insight into Saddam's thinking. Postwar interviews with his top commanders suggested Saddam was also thinking about a replay of 1991, but he intended to repeat his 1991 "success" in staying in power.[49] Saddam was prepared to endure a heavy bombing campaign, just as he had before. But he was counting on the Americans being too fearful of casualties to come all the way to Baghdad. Until almost the very end he was more fearful of another Shiite revolt or intervention by Iran than of US troops.[50] There was, of course, as much fantasy as reason in these calculations. But that is one more reason why strategic bombing has a hard time changing minds.

Numerous other regime leaders were also targeted in the initial air strikes, but none were killed. Precision can not make up for extremely faulty intelligence. (Special bunker-piercing bombs were used on the site where Saddam was supposed to be, against a bunker that did not exist.)[51]

As for the ground campaign, the impact of the battlefield bombing was significant without being dramatic. The postwar interviews showed that even dropping leaflets on the Republican Guard troops was enough to begin the defections.[52] It took only a small amount of actual PGMs for the defections to grow to 50 percent or more of total troop strength.

Indeed there wasn't even room for classic close air support, because there were few formal engagements with the regular forces.[53] Most of the fighting was with irregular forces armed with RPGs, machine guns, and pickup trucks. Some air strikes were used during these confrontations. For both the Army and the Marines, a JDAM attack provided new flexibility and sometimes more accuracy than traditional artillery, but urban warfare is hard to conduct from the air.

Of course the evolution of the conflict into a counterinsurgency campaign has stressed the doctrines of all the services. Airpower of any sort has much more constrained utility under those conditions.

Technology and organization for future conflicts: CAOC and UAVS

As with precision weapons, most of the technologies favored by net-centric advocates also enhance Air Force core interests, and thus have been supported by the service. As they get developed and deployed, these technologies also have a political dimension. They allow the Air Force to play the role of experienced senior partner when joint, interoperable, cross-service programs are implemented.

In the net-centric future envisioned by its advocates, all sensors and communications channels would be seamlessly and fully integrated so that battlefield situational awareness would be available as needed to commanders up and down the chain of command. In theory, local commanders could then use the information to act on their own initiative, with a great deal of decentralization ("self-synchronization").

But the path to that future requires enormous centralization to integrate all those sensors and communications systems. This centralization, in turn, also allows for a much tighter command and control system, which matches the Air Force's historic preferences.

During World War II, when the Air Corps was still working for organizational independence, it strongly argued that airpower should be centrally controlled by Air Corps officers. For only they understood the impact of massed airpower, in contrast to ground force officers, who might lose the advantage on isolated efforts to support ground troops. The Air Force sometimes won this fight during World War II and Korea, but often was rebuffed. During Vietnam, the Navy air campaign was largely independent of the Air Force.

But the Goldwater-Nichols Act strengthened the powers of the combatant commanders to run integrated field campaigns. For Desert Storm, General Schwarzkopf similarly agreed that there would be one joint air component commander to oversee all Air Force, Navy, Marine, and Coalition aircraft, in a unified campaign. A single joint tasking order was created for each day, specifying the roles of bombers, fighters, tankers, recon aircraft, etc.

However, it took a staff of thousands in the Air Operations Center to plan the 500 to 1,000 sorties in each daily order. So, in reaction to the Desert Storm experience, the Air Force in 1995 laid out a plan to institutionalize and computerize the Centers. The aim was both to automate the analysis and planning that went into the creating of the Air Tasking Order itself, and to attempt to fully integrate the various reconnaissance and communications systems to provide real-time battlefield situational awareness. Five permanent regional mega-centers, labeled Combined Air Operations Centers (CAOCs), were to be created around the world.[54]

In 2000, General John Jumper, then Air Force Chief of Staff, indicated his support for the CAOCs by making their development a formal Air Force weapons system.[55] He had observed how the skeleton CAOC that was available at the beginning of the Kosovo campaign was a very inadequate starting point for a short-notice operation.[56] This new status gave the program its own project manager, budget line, and dedicated training slots. The goal was to develop standardized technologies, procedures, and staffing – managing air operations would become a specialty on par with existing Air Force slots.

Operation Iraqi Freedom was again fought as a joint air campaign, managed from the Saudi Arabian CAOC. It demonstrated both the clear gains from Desert Storm and the long path to full net-centric capabilities. In addition to coordinating the strategic bombing campaign, the controllers also were able to more effectively provide close air support. The failure of the Scud hunt during Desert Storm had forced more attention to "time-sensitive targeting," and the tighter recon/communications loops that had been developed, combined with PGMs, extended to all the ground forces.

There was also a closer integration between the ATO and the parallel tasking order for space assets. Satellite communication capacity, satellite sensor availability and GPS satellites were matched to the rest of the campaign. (GPS signals were tweaked each day to get maximum accuracy when bombing was starting.)[57]

With the departure of US troops from Saudi Arabia in August 2003, the Prince Sultan CAOC was closed and control of air operations in both Iraq and Afghanistan was shifted to a new CAOC at Al Udeid Air Base in Qatar. Yet the task of full scale integration still has far to go: "Hundreds of target planners, intelligence analysts, weather forecasters and other specialists at an air operations center run more than fifty different communications and computer systems, many of which cannot exchange information digitally" (Sandra I. Erwin, "Memo to War Planners: No Post-It Notes or Stubby Pencils Allowed").[58]

There are some inherent problems with the technological vision that the Air Force is aiming for. First, many of the individual technologies have operational flaws. Even a sophisticated system like JSTARS can lose its main radar coverage one or more times in a twelve-hour shift. Second, the data fusion is years or decades away from the seamless operation that is promised. In addition, as the technologies do become meshed, the network itself becomes vulnerable to accident or attack.

Future paths – organizational

Air Force commanders have undertaken the CAOCs as an Air Force independent program, but they are happy to discuss the integration of Army or Navy data as well, with the unspoken assumption that because they have gone the farthest in this area, the Air Force systems should be the centerpiece of any joint system of systems. Obvious to all was that Air Force officers expected to command CAOCs and thus the airwar in all of its aspects.

It is the Navy that probably has the most to lose organizationally. About half the sorties in Operation Iraqi Freedom that were jointly controlled were flown by Navy and Marine aircraft. As the CAOC command and control becomes tighter and more effective, the room for an independent naval air role (at least for ground targets) grows smaller. The Navy has tried to adapt to this by playing a more active role in the CAOC. The deputy commander is now usually from the Navy and each of the carrier groups involved has a liaison team on-site. There is much more cooperation on using Air Force (and sometimes British) tankers to extend the range of Navy strike planes. But they are still the junior partner.

For the Army the changes are more mixed. Close air support with precision weapons can be a very effective adjunct. Airborne surveillance also holds great promise. For the Army, the key is how the information is actually distributed to field commanders. During the invasion in 2003, frontline commanders felt that little information about enemy dispositions was reaching them, and some information that was supplied (about the location of key Republican Guard divisions) was wrong or out of date.[59] On the other hand, more recently, Army officers aboard Air Force JSTARS planes are giving orders to Army convoys about where to look for IEDs.

One aspect of the new technology will not be favored by any of the services, but will be hard to avoid. That is, more effective real-time command and control of an air campaign also offers the prospect of more intervention from civilian higher authorities. The Air Force already dislikes civilian intervention in target selection (particularly during the Kosovo campaign, when it involved not just Washington, but Brussels). But the more seamless the system, the easier it becomes to weigh in from any part of the network. Both precision munitions and networks raise expectations among civilian leaders.

Developing and managing UAVs

"UAVs are air vehicles, after all, and their development and employment will benefit from guidance from the masters of airpower" (Rebecca Grant, *Air Force Magazine*).[60] In fifty years of involvement with Unmanned Aerial Vehicles (UAVs), the Air Force has gone from over-enthusiastic to uninterested to wanting to control them all. Indeed they now think the Combined Operations Center is exactly the proper home for UAV management and control.

In fact all of the services, plus the CIA and the National Reconnaissance Office, had UAV projects at some time, but with little success until the general technological advances of the last decade.[61] UAVs long held out the promise of inexpensive, flexible access to enemy territory. But the available technologies did not allow for accurate navigation, or high enough data flow, or even effective control of the aircraft.[62]

Only with the advent of GPS for effective navigation and wide-bandwidth satellite communications to transmit the surveillance pictures did UAVs become of military value. The Predator, the first successful low-level UAV, was an Israeli/General Atomics design, fielded in Bosnia in 1993–1995 by CIA and Army teams.[63]

At this point, General Fogelman, then the Air Force Chief of Staff, personally intervened. He could see a "transition point" had been reached. Because unmanned aircraft were here to stay, they should be flown by the Air Force, despite its long neglect. He won support for the idea within his service and then waged a year-long battle until Predator management and future development was established as an Air Force line item.

However the success of UAVs in Afghanistan and Iraq has inspired all the services to develop their own models. The Army has contracted with General Atomics for Sky Warrior, its own version of the Predator. In the summer of 2007, the Navy selected Northrup Grumman to develop a carrier-capable UAV, much like the Air Force's high-flying Global Hawk.[64]

Senior Air Force commanders became uncomfortable with this competitive pluralism. During 2006–2007 they became increasingly determined that the Air Force should become the "executive agent" for all UAVs that fly above 3,500 feet. They argue that UAVs are the next wave of the net-centric universe and need to be developed jointly. Furthermore they should be operationally managed theater-wide by the joint air commander, through the CAOC.[65] Lt. General Deptula, now an Air Force Deputy Chief of Staff, testified before a House Committee on the urgency of this plan in April, 2007.

Because the Army is developing its UAV specifically so it can be organically attached to each brigade or other unit, this proposal has been very controversial. Indeed, all the other services are opposed.[66] But Congressional committees and some DoD civilian managers are again attracted to the supposed efficiencies of jointness.

It would be far better if Congress learned from the long history of these programs. The renewed Air Force interest in UAVs only came about because of the prior interests of other agencies. As described in the Cote/Sapolsky model (see Chapter 1, p. 8), bureaucratic competition was the spur to innovation. However, this is still an emerging technology, with room for far greater innovations that will provide many new capabilities. It is also one where decentralized deployment is well matched to the nature of the technology.

Thus, instead of trying to suppress competition, this is a technology area well suited to multiple competitive efforts from different agencies. Congress and the DoD should try to structure the competition so that the innovative designs that best meet the program goals receive additional funding, and so that the competing programs are encouraged to learn from each other. Structured competition, not jointness, is most likely to lead to effective, responsive innovation.

In September 2007, Deputy Defense Secretary Gordon England refused the Air Force's request, opting for a series of joint oversight committees. However he also ordered that the Predator and Sky Warrior programs, the two most similar, should be merged.[67] Air Force Chief of Staff General Michael Moseley saw this not as a defeat, but as one more step toward central control, always the Air Force preference.[68]

The Air Force has similarly argued that it should develop and manage all national security space-based assets. It wants to take control of not only the satellites now run by the Army and Navy, but those from the intelligence community as well.[69]

By the summer of 2008 though, the tides had shifted, as Robert Gates, the new Secretary of Defense, began to assert himself. General Moseley was fired, and replaced as Chief of Staff by the first non-fighter pilot in that post in more than twenty years.[70] Gates publically decried the lack of interest from the Air Force in getting more UAVs to the battlefield in Iraq where they could serve Army requests. He spurred the Army to field its own combinations of UAVs and attack helicopters.[71] Thus Gates is trying to induce some interservice competition from the top down, but he is dependent on the support of the new administration.

Conclusions

Twenty years of technological opportunities for the Air Force have produced considerable "creation" but little or no "destruction." Capabilities have greatly increased, but there has been no corresponding shift in the overall force structure. This is not surprising. The new capabilities allowed the Air Force to do traditional missions more effectively, and were thus welcomed. A larger shift would have required a more determined "buyer" at the Department of Defense level. Indeed, the Air Force perspective on the appropriate reallocation would be a smaller Army.

There has been no acknowledgment that expensive programs such as the F-22 have little place in the current threat environment. In fact, the Air Force is going in the opposite direction. The F-22 is being retrofitted so that it too can be a platform for precision bombing. What cutbacks that have occurred have been mostly incremental. They have taken place within the context of the annual budget negotiations with Congress and the DoD. These budget constraints have been largely separate from the process of innovation.

However, the Air Force is still planning for at least 183 F-22s for $65.3 billion (down from 740 in 1990).[72] A new, super-stealthy heavy bomber is being sought by 2018.[73] The major tanker replacement program is overdue. And the services together want 2,500 Joint Strike Fighters (at least 1,753 F-35As or Bs for the Air Force), which would cost an astonishing $300 billion.

The Air Force finds itself increasingly mired in a collection of ironies. After nearly 100 years of promises, strategic bombing finally has the precision, stealth, and firepower to do enormous damage to an enemy's military, economy, and society. Yet, with the exception of the Milosevic experience, this power does not seem as coercive as intended. Second, attempts to target regime leaders personally require better and more timely intelligence than we have been able to produce.

The very precision desired for strategic effects has proven more useful for battlefield interdiction and close air support (even though these were always secondary missions for Air Force planners). Any formal military force that wants to go head to head with US forces risks being smashed like the Republican Guard of 1991. But, just as the Air Force has grown to appreciate joint efforts with the ground forces, the opportunities are diminishing. First in Kosovo, and now with the full-scale Iraqi insurgency, our enemies have learned very small units are the way to fight.

Air Force Major General Charles Dunlop has recently noted the expanding use of precision airpower against even insurgent targets.[74] The growing technological capabilities of real-time surveillance and smaller bombs that impact a limited area have created incremental changes in tactics effective for village level firefights. But, as Dunlop points out, these up-from-the-ranks adaptations find little support to date in either Air Force or Army official doctrines. It remains to be seen if they get enough sustained support to become full-scale innovations.

As noted above, one reaction to the successful development of stealth technology would have been to create a specialized cohort of planes that used stealth in the initial stages of an attack on enemy defenses, until that air defense system had been put out of action. Much less costly fighters could then take up the battle.

The Air Force has chosen to go in the opposite direction, and make every weapon stealthy, at enormous cost. All these stealthy platforms, which now all carry precision weapons, vastly increase the effective striking power per sortie. Again, the Air Force could have responded to these greatly increased capabilities with a new doctrine which stressed a much reduced but still highly capable force (beyond the reductions forced by budget constraints). Instead the Air Force has given new life to a ninety-year-old doctrine of warfighting that calls for endless aim points, but never delivers the promised impacts. Clearly the new technologies have been good for the Air Force, but no transformation was desired or achieved.

Although the technologies now on order are questionable responses to likely foreign enemies, they are proving very useful in the Air Force's domestic battles with the other services. Expertise in communications, surveillance, and coordination technologies does give the Air Force a leading position in the Pentagon debates about interoperability and joint operations. That is a battle where they may find success. Whether new military and civilian leadership at the top will result in different priorities for the Air Force remains to be seen.

Notes

1 Michael Brown, *Flying Blind: The Politics of the US Strategic Bomber Program* (Ithaca, NY: Cornell University Press, 1992), p. 38.
2 See the fine literature summary in Brown, *Flying Blind*, Chapter 2.
3 Donald L. Miller, *Masters of the Air: America's Bomber Boys Who Fought the Air War Against Nazi Germany* (New York, NY: Simon & Schuster, 2006), Chapter 1; Barry D. Watts, *The Foundations of US Air Doctrine: The Problem of Friction in War* (Maxwell Air Force Base, AL: Air University Press, 1984), Chapter 3.
4 Miller, *Masters of the Air*, Chapter 16.
5 Richard Hallion, "Precision-Guided Munitions and the New Era of Warfare," *Air Power History*, Vol. 43, No. 3 (Fall 1996), p. 7.
6 Hallion in Bernard C. Nalty (ed.), *Winged Shield, Winged Sword: A History of the United States Air Force* (Washington, DC: Air Force History and Museums Program, United States Air Force, US GPO, 1997).
7 Quoted in Richard Hallion, *Storm Over Iraq: Air Power and the Gulf War* (Washington, DC: Smithsonian Institution Press, 1999), p. 209.
8 John Warden, *The Air Campaign: Planning for Combat* (Washington, DC: National Defense University Press, 1990).
9 Michael R. Gordon and Bernard E. Trainor, *The Generals' War: The Inside Story of the Conflict in the Gulf* (Boston, MA: Little, Brown, 1995), p. 77.
10 Gordon and Trainor, *The Generals' War*, pp. 93–96.
11 Ibid., p. 474.
12 For similar views see Rick Atkinson, *Crusade: The Untold Story of the Persian Gulf War* (Boston, MA: Houghton Mifflin, 1993); Hallion, *Storm Over Iraq;* and Benjamin S. Lambeth, *The Transformation of American Air Power* (Ithaca, NY: Cornell University Press, 2000). The official Air Force study is Eliot Cohen (ed.), *Gulf War Air Power Survey* (Washington, DC: US Air Force, 1993).
13 Robert Pape, *Bombing to Win: Air Power and Coercion in War* (Ithaca, NY: Cornell University Press, 1996). The most stinging attack on the airpower theories, arguing that it was actually vastly superior ground technology and training that won the victory, is Daryl G. Press "The Myth of Air Power in the Persian Gulf War and the Future of Warfare," *International Security*, Vol. 26 (Fall 2001), pp. 5–44.
14 Warden left the service without making general, but his fellow planner, David Deptula, now is a senior member of the Air Staff.
15 Brown, *Flying Blind*, Chapter 7.
16 Marshall L. Michel, *Clashes: Air Combat over North Vietnam, 1965–1972* (Annapolis, MD: Naval Institute Press, 1997), pp. 10–13; see also C. R. Anderegg, *Sierra Hotel: Flying Air Force Fighters in the Decade after Vietnam* (Washington, DC: Air Force History and Museums Program, US Air Force, US GPO, 2001).
17 R. Michael Worden, *Rise of the Fighter Generals: The Problem of Air Force Leadership, 1945–1982* (Maxwell Air Force Base, AL: Air University Press, 1998), Chapter 8.
18 Kenneth P. Werrell, *Chasing the Silver Bullet: US Air Force Weapons Development from Vietnam to Desert Storm* (Washington, DC: Smithsonian Books, 2003), Chapters 3–4.

19 Stephen Hill, *Weapons Innovation and Joint Systems Development: a Case Study of the Joint Stars Program* (Masters thesis, Cambridge, MA: MIT, June 1997).
20 Michael Rip and James M. Hasik, *The Precision Revolution: GPS and the Future of Aerial Warfare* (Annapolis, MD: Naval Institute Press, 2002), pp. 233–238.
21 George Cahlink, "Birth of a Bomb," *Government Executive*, August 2003, p. 23.
22 Mark A. Lorell and John C. Graser, *An Overview of Acquisition Reform Cost Savings Estimates* (Santa Monica, CA: Rand, 2001), Chapter 4.
23 Ibid.
24 Bill Sweetman, "The Falling Price of Precision," *Jane's International Review* (April 2002), p. 47.
25 Cahlink, "Birth of a Bomb," p. 26.
26 Keegan, *Electronic Telegraph*, June 6, 1999, cited in Hallion, *Storm over Iraq*, p. ix.
27 Hallion, *Storm Over Iraq*, pp. viii–ix.
28 The Air Force also calls its deployment groups "expeditionary forces." These are integrated packages of planes, airmen, and support units that train together and then deploy to the combat theater for 120-day tours. Then they are rotated out and replaced with another group. While this development has eased the logistical strain of a long war, it has little to do with engaging the enemy. See John Tirpack, "On Expedition, Hanging Tough," *Air Force Magazine* (September 2007), pp. 32–37.
29 Andrew J. Bacevich and Eliot A. Cohen (eds.), *War over Kosovo: Politics and Strategy in a Global Age* (New York, NY: Columbia University Press, 2001); Stephen D. Wrage (ed.), *Immaculate Warfare: Participants Reflect on the Air Campaigns Over Kosovo, Afghanistan, and Iraq* (Westport, CT: Praeger, 2003).
30 Benjamin S. Lambeth, *NATO's Air War for Kosovo: A Strategic and Operational Assessment* (Santa Monica, CA: RAND, 2001), Chapter 7.
31 Lambeth, *NATO's Air War for Kosovo*, pp. 27–28.
32 Lambeth, *NATO's Air War for Kosovo*, Chapter 6.
33 Scott A. Cooper, "Air Power and the Coercive Use of Force," in Wrage, *Immaculate Warfare*, p. 12.
34 William Arkin, "Operation Allied Force: 'The Most Precise Application of Air Power in History,'" in Bacevich and Cohen, *War over Kosovo.*
35 Lambeth, *NATO's Air War for Kosovo*, Chapter 6.
36 Lambeth, *NATO's Air War for Kosovo*, pp. 38–43.
37 Arkin, "Operation Allied Force," p. 26.
38 Short, quoted in Scott A. Cooper, "The Politics of Air Strikes," in Wrage, *Immaculate Warfare*, p. 73.
39 For a summary of the large literature on this point see Bruce R. Nardulli, *Disjointed War: Military Operations in Kosovo, 1999* (Santa Monica, CA: Rand, 2002), pp. 41–42; and Arkin, "Operation Allied Force."
40 Cooper, "The Politics of Air Strikes."
41 Benjamin Lambeth, *Air Power Against Terror: America's Conduct of Operation Enduring Freedom* (Santa Monica, CA: RAND, 2005), p. 248.
42 Benjamin S. Lambeth, *American Carrier Air Power at the Dawn of a New Century* (Santa Monica, CA: RAND, National Defense Research Institute, 2005), p. 28.
43 Rebecca Grant, "The Carrier Myth," *Air Force Magazine* (March 1999).
44 Owen R. Cote, *The Future of Naval Aviation* (Cambridge, MA: Massachusetts Institute of Technology, Security Studies Program, 2006), pp. 53–67.
45 Cited in John T. Correll, "Basic Beliefs," *Air Force Magazine* (June 2004), pp. 45–46.
46 Michael R. Gordon and Bernard E. Trainor, *Cobra II: The Inside Story of the Invasion and Occupation of Iraq* (New York, NY: Pantheon Books, 2006), p. 89.
47 Gordon and Trainor, *Cobra II*, pp. 169–177.
48 Gordon and Trainor, *Cobra II*, p. 209.
49 Kevin M. Woods with Michael R. Pease, Mark E. Stout, Williamson Murray, and James G. Lacey, *Iraqi Perspectives Project: A View of Operation Iraqi Freedom from*

Saddam's Senior Leadership (Joint Center for Operational Analysis, Joint Forces Command, 2006), Chapter 2.

50 Ibid.

51 Gordon and Trainor, *Cobra II*, p. 209.

52 Woods *et al.*, *Iraqi Perspectives Project*, p. 125.

53 Gordon and Trainor, *Cobra II*, Chapter 20.

54 Prince Sultan Air Base, Saudi Arabia; Osan Air Base, South Korea; Hickam Air Base, HI; Davis-Monthan Air Force Base, AZ; and Ramstein Air Base, Germany. In addition, an experimental CAOC was set up at Langley to test out new technology before it was deployed at the operating centers.

55 Lambeth, *NATO's Air War for Kosovo*, p. 202.

56 Ibid.

57 Adam J. Hebert, "Toward Supremacy in Space," *Air Force Magazine* (January 2005).

58 *National Defense* (October 2004).

59 Gordon and Trainor, *Cobra II*, Chapter 19.

60 Rebecca Grant, "The Clash of the UAV Tribes," *Air Force Magazine* (September 2005).

61 Albert D. Wheelon, "CORONA: a triumph of American technology." In Dwayne A. Day, John M. Logsdon, and Brian Latell (eds.), *Eye in the Sky: The Story of the Corona Spy Satellites* (Washington, DC: Smithsonian Institution Press, 1998).

62 Thomas Ehrhard, *Unmanned Aerial Vehicles in the United States Armed Services: A Comparative Study of Weapon System Innovation* (PhD Thesis, Washington, DC: The Johns Hopkins University, June 2000), Chapter 1.

63 Laurence R. Newcome, *Unmanned Aviation: A Brief History of Unmanned Aerial Vehicles* (Reston, VA: American Institute of Aeronautics and Astronautics, Inc., 2004).

64 Robert Brodsky, "Navy Plans to Land Pilotless Drones on Carriers," *Government Executive*, August 10, 2007.

65 Rebecca Grant, "The Drone War," *Air Force Magazine* (July 2007).

66 Demetri Sevastopulo, "US Military in Dogfight Over Drones," *Financial Times*, August 19, 2007; Megan Scully, "Air Force Still Pressing to Control Unmanned Aerial Vehicle Programs,"*CongressDaily*, June 21, 2007.

67 Megan Scully, "Pentagon Rejects Air Force Bid to Control UAV Programs," *CongressDaily*, September 14, 2007.

68 *Air Force Magazine Online*, "Daily Report," September 25, 2007.

69 Andy Pasztor, "Air Force Seeks Greater Control Over Military And Spy Satellites," *Wall Street Journal*, November 7, 2007.

70 Secretary of the Air Force Wynne was also fired. The official reason was deficiencies in the Air Forces stewardship of its nuclear weapons, but it was clear that Gates had lost confidence in them in many areas. "Moseley and Wynne Forced Out," *Air Force Times*, June 9, 2008.

71 Thom Shanker, "At Odds With Air Force, Army Adds Its Own Aviation Unit," *New York Times*, June 22, 2008.

72 John Tirpack, "The Air Force Starts Over," *Air Force Magazine* (August 2007).

73 Adam Hebert, "Great Expectations," *Air Force Magazine* (August 2007); and Megan Scully "Air Force Envisions Stealthy Long-range Bomber for 2018," *CongressDaily*, June 13, 2007.

74 Charles J. Dunlap, Jr., "Making Revolutionary Change: Airpower in COIN Today, *Parameters* (Summer 2008), pp. 52–66.

6 The Marine Corps

Sticking to its guns

Austin Long

The Marine Corps' experience in the second interwar period was marked neither by the angst felt in the Army nor by the technological epiphanies that occurred to some in the Air Force and Navy. Nor did it resemble the Marine experience in the first interwar period, which was marked by both doctrinal and technical innovation as well as bitter factional disputes inside the Corps.[1] Instead, the Marine Corps found itself in a relatively good position to continue performing its traditional mission of expeditionary warfare.

Two opportunities for major transformation were missed in the second interwar period. First, the technological changes and doctrinal debates started in the Cold War era might have culminated in a net-centric doctrine that revolutionized marine expeditionary warfare. This vision was partially evident in the plans for "Ship to Objective Maneuver" and related initiatives, but was stillborn in its execution: no real conventional transformation occurred.

Second, an environmental transformation propelled by the small wars of the 1990s could have pushed the small wars community in the marines to develop a new doctrine of counter-insurgency suitable to the future of warfare. While some small technological and doctrinal changes did occur in this area, by and large this transformation was missing as well – the maneuver warfare community in the marines usually controlled the pace of marine politics.

These absent changes have many fathers. Most notably, the variables suggested by Stephen Rosen's theory of intra-service competition seem to have taken on values unfriendly to transformation. The culture of the Marines appears to have prevented the continuity in leadership that can give potential innovators protection. Those small changes that did occur were influenced by the inter-service position of the Corps. Acutely attuned to the threats to its autonomy from other services, the Marine Corps was willing to make small changes that bolstered its political position.

However, the Corps was not entirely static in this second interwar period. This chapter will discuss the changes that did take place. It proceeds in five parts. The first section briefly discusses some of the idiosyncrasies of the Marine Corps as an organization. The second provides a short summary of what was going in the Marine Corps in the period prior to the second interwar period (from the end of Vietnam to the end of the Cold War) as a necessary backdrop. Much of the Marine experience in the second interwar period is a continuation of trends begun in this era. The third section focuses on the early interwar period, from Desert Storm to the intervention in Haiti. The fourth section examines the later interwar period, as debates about an emerging "revolution in military affairs" (RMA) began to gather momentum even as defense budgets remained somewhat tight. The fifth section discusses the Marine experience in the early part of the new conflict some now call "the Long War."

The Marine Corps as an organization

The modern Marine Corps is an odd organization in a number of ways. It has no truly unique mission to dominate a medium (i.e. land, sea, or air) like the other three services, yet it functions in all three. It views itself as an elite infantry ("every man a rifleman" is still taken quite seriously), yet it includes fighter pilots, tank drivers, and other non-infantry. Its members agree that they are "expeditionary warriors," but some think this mostly means conducting amphibious assaults while others think it means fighting small wars. These two communities compete for position and influence within the organization, often trading off the position of Commandant. The Marine Corps' existence and even a floor on its force structure (a minimum of three air wings and three divisions) have been enshrined in law for six decades, yet it remains the most paranoid of the services about being dissolved or absorbed into another service.[2]

It is this latter characteristic that had the most effect on the Marine Corps in the second interwar period. Despite being in a relatively good position bureaucratically, the Marine Corps leadership is always willing to at least appear to embrace new missions, particularly if those missions are not very different from what the Marine Corps is already doing or if it can be done relatively cheaply. Marines can then claim that this "flexibility and adaptability" is a hallmark of the Corps and a major argument for its continued existence.[3]

This bureaucratic insecurity has, over time, produced a second element of Marine distinctiveness. The Marine Corps is probably the most politically savvy of the services, perhaps rivaled only by the Air Force. It goes out of its way to cultivate support from both elite groups and the broader US population by holding such events as its Capabilities Exercise (CAPEX). CAPEX, essentially a high-tech dog-and-pony show staged for political benefit, takes place at Camp Lejeune and Camp Pendleton for an audience of invited guests ranging from the media to civilian defense analysts to members of various civic groups. It includes displays of Marine capabilities and equipment that are explicitly "designed to demonstrate the flexibility and combat power of the Marine air-ground task force."[4]

The Marine Corps is also sensitive to the importance of public affairs and Congressional liaison. Congress has been critical to the preservation and advancement of the Marine Corps and its interests, and the Marines maintain close relations with many Congressmen.

In short, the Marine Corps is an organization with two important structural features: it is divided into two communities who vie for supremacy in the organization, while the organization itself is very sensitive to competition from other services for its job. This makes for a potentially interesting environment from the view of innovation theory.

From Vietnam to the end of the Cold War

The Marines, who had been the first combat units into Vietnam, were the first service to be fully withdrawn from the conflict, with the last Marine units leaving in 1970. It thus had something of a headstart in planning for the future while the other services remained engaged. Further, unlike the other services, the Marine Corps only briefly had to rely on the draft during World War II and Vietnam, so the subsequent shift to an all-volunteer force probably impacted it the least. These two advantages meant that the Marine Corps should have had an easier time than the other services in adjusting to the end of the war.

However, the post-Vietnam period was difficult for the Marine Corps. Defense budgets were tight. The American appetite for the small wars that many in the Marine Corps saw as its special competence had reached its nadir. The country was refocusing on the threat posed by the Soviets, particularly to Western Europe. All three of the other services were focused on their contributions to NATO in a future war. The Marines, bureaucratically insecure and knowing which way the wind was blowing, reacted similarly.

Marine interest in NATO missions was not entirely new. Large exercises in the Mediterranean dated to the 1960s. However, the line of thinking that developed in the 1970s, largely by Navy advocates, provided a new opportunity.[5] The idea was that in the event of a conventional clash between NATO and the Warsaw Pact, NATO's "northern flank" would need additional protection, particularly Norway's far north. The Marine Corps leapt at the opportunity. In 1975, a Marine Amphibious Unit conducted maneuvers in Norway and Germany, the Marines' first NATO operation away from the Mediterranean. By the end of the 1970s, brigade-sized maneuvers were being conducted by the Marines in Norway.[6]

These maneuvers led to substantial Marine investment in equipment prepositioning. This concept was being widely debated in the 1970s, and was eventually adopted by the Army as the "Prepositioned Overseas Material Configured to Unit Sets" program in Europe. Many in the Marines felt that prepositioned material undercut the unique expeditionary nature of the Marine Corps and could eventually be used as a bureaucratic argument by the Army against the Corps. But the proponents of prepositioning won this debate, and Norway became home to the only land-based Marine prepositioned equipment.[7] This operational construct, eventually known as the Norway Air-Landed Marine Expeditionary Brigade (NALMEB), meant that a Marine brigade would fly into Norway, marry up with equipment stored in climate-controlled caves and tunnels, and then deploy throughout Norway.[8]

Even as NALMEB was being developed to fill a perceived need in national strategy, three other debates were underway. One of the most notable was on the need for a Rapid Deployment Force (RDF) to quickly move to reinforce states threatened by Soviet expansion. By the late 1970s, the principal area of concern was the Middle East, as Soviet naval power in the Mediterranean appeared to be increasing even as the Soviets moved into Afghanistan to prop up their client state. The United States had lost a primary ally in the Persian Gulf with the ouster of the Shah in 1979.

Some in the Marine Corps, such as General P. X. Kelly, embraced the RDF concept as another mission for which the Corps was well suited.[9] However, confronting a Soviet Motor Rifle Division was going to be challenging for a force that was traditionally organized for either a relatively brief amphibious assault or for more low-intensity conflict. Like the northern flank, Marine participation in the RDF would lead to significant internal debates and subsequent changes in the Corps.

The first debate was again about prepositioning. In the case of the RDF, this could not be done on land, in large part because the United States did not have the same formal alliances with Gulf States that it had with NATO members like Norway. Instead, Maritime Prepositioning Ships (MPS) were developed and deployed to Diego Garcia in the Indian Ocean beginning in 1980. The same concerns about prepositioning that emerged in the NALMEB debate emerged in the RDF debate. The Army was also going to deploy its own prepositioning ships. If the Army could provide exactly the same capability, some argued, there was no advantage for the Marine Corps in devoting scarce resources to the problem.[10]

The second debate was about force structure and acquisition. Some felt that to deal with both NATO and RDF contingencies, the Marine Corps should acquire systems that would enable Marine units to engage Soviet armored forces. An example was acquiring the Army's new M1 main battle tank.[11] Others felt the Marines should avoid getting too heavy in terms of equipment, as it undermined their ability to be expeditionary. Further, some argued that the Marines should produce specialized units for the Norway and RDF contingencies.[12]

The final debate was doctrinal. The increasing emphasis on engaging Soviet armored or mechanized forces meant that the Marines would be engaging in warfare outside the mainstream of traditional Marine doctrine. In answer to this challenge, the Marine Corps adopted a new doctrine termed "maneuver warfare." Essentially a combination of German concepts such as *auftragstaktik* (directive command) with Air Force Col. John Boyd's theory of the OODA loop, maneuver warfare was adopted as part of the new Marine Corps capstone doctrine in the 1989 manual FMFM 1 *Warfighting*. Maneuver warfare centered on pushing decision-making to the lowest levels of command within a fairly broad "commander's intent."[13]

Finally, even as these debates took place, the Marine Corps continued to be the force of choice for small-scale US interventions. Most notable was the intervention in Lebanon in the early 1980s. This operation ended tragically following the bombing of the Marine barracks in Beirut, which killed over 200 marines.

The second interwar period begins

The end of the Cold War came with debates from the post-Vietnam era in full tilt. On one side were advocates of the heavier, maneuver warfare doctrine and force structure.[14] These analysts felt that the Marines, with their organic fixed-wing aviation and skill at close air support, were in a strong position to execute the kind of truly integrated, all-arms operation that was the hallmark of the German blitzkrieg and the goal of the US Army doctrine of AirLand Battle.

On the other side of the debate were advocates of Marine competence in "small wars" or "low-intensity conflict." They argued that low-intensity conflict was the form of warfare that the Marines were most likely to be called on to wage in the future, and it should be the focus of preparation and doctrine.[15] Rather than getting heavier, the Marines should stay relatively light and avoid heavy reliance of prepositioning and the massive logistical requirements that were the hallmark of the Army.

The two Commandants of the late Cold War and just after it embodied this debate. P. X. Kelley, Commandant from 1983 to 1987, was a major proponent of the maneuver warfare doctrine, which was developed during his tenure. His successor, Alfred Gray, Commandant from 1987 to 1991, focused more on low-intensity conflict and took steps to move the Corps away from the heavy units embraced by his predecessor. For example, Gray decided to cut back heavily on Marine acquisition of the M1 tank.[16]

However, just as Gray seemed in position to shift the Marine Corps away from maneuver warfare, Saddam Hussein's Iraq invaded neighboring Kuwait. In just a few months, the advocates of prepositioning and maneuver warfare saw their concepts validated as the 1st and 2nd Marine Divisions surged into Saudi Arabia as part of Operation Desert Shield and Operation Desert Storm. To many, the short, intense, and successful Gulf War – not low-intensity conflict – was the model for the Marines to emulate in the future.

One immediate impact of the war was the decision to acquire heavier and more modern armor. The 2nd Marine Division was already equipped with M1 tanks, but the 1st Marine Division still retained M60A1 tanks. Though these tanks performed well against the poorly trained Iraqis in Desert Storm (for example, at the Battle of al-Khafji), it seemed unwise to maintain the M60 for use against more competent foes. By the end of 1991, the M60 was phased out of Marine service, and the Marines acquired additional M1A1s, state-of-the-art at the time. From a low of somewhat over 200 M1s, the Marine inventory grew to slightly over 400 M1A1s. About 180 of the 400 were transferred from the Army rather than purchased (some Army officers might say "stolen").[17]

Desert Storm was also the combat debut of another system that the Marine Corps had been experimenting with under both Kelley and Gray: the Pioneer unmanned aerial vehicle (UAV). Despite their disagreements about likely missions, both commandants felt that a lightweight and expendable reconnaissance asset like Pioneer would be of use to the Marine Corps. Following Desert Storm, Gray helped sponsor other UAV projects like the very small Pointer UAV system.[18]

Even as the Marine Corps got heavier following Desert Storm, it also began to embrace two other, related challenges. The first was a focus on urban combat and operations. The second was the need for nonlethal munitions to deal with military operations other than war (MOOTW).

The urban combat imperative became readily apparent in 1993 following the Battle of Mogadishu, immortalized in the book *Black Hawk Down*.[19] There, large numbers of irregular opponents from the Habr Gidr clan engaged well-armed US special operations forces (SOF) on the streets of Mogadishu, Somalia. Though they acquitted themselves well, the SOF forces were bloodied. The apparent difficulty of fighting in urban terrain for even well-trained forces made the need for better preparation evident. Other contemporaneous events, such as the disastrous first Russian attempt to invade Grozny in 1994, further highlighted the need to prepare for urban combat.[20]

To some analysts, the Battle of Mogadishu also illustrated the need for nonlethal weapons. Many civilians had been caught in the crossfire. After overseeing the withdrawal from Somalia known as Operation United Shield, Marine Lt. Gen. Anthony Zinni asked for a quick response capability to deploy improved nonlethal weapons.[21] This request had little immediate impact, but it accelerated nonlethal weapons development efforts by Marines.

Operation Uphold Democracy, an intervention in Haiti that began in 1994 and continued in various forms for two years, reinforced the desire for nonlethal weapons. The operation involved significant interaction with a potentially violent local population. Some Marines (and soldiers) who participated in Uphold Democracy felt that nonlethal weapons provided a useful interim step between no reaction and lethal force.[22]

Commandant Charles Krulak's response to these challenges was to set up the Commandant's Warfighting Laboratory (henceforth, the Lab) in 1995.[23] Commandant from 1995 to 1999, Krulak was sympathetic to the small wars school of thought. Remarking on the future of war, he famously noted, "We're worried about the son of Desert Storm when what we are seeing is the stepchild of Chechnya."[24] He intended for the Lab to explore new technological and doctrinal concepts for the Marine Corps with emphasis on urban warfare and nonlethal weapons. Renamed the Marine Corps Warfighting Laboratory in 1997, the Lab invested heavily in technology for the urban fight, including systems that would later fall under the rubric of "transformation," such as information-sharing systems. It also spent time experimenting with new tactics, techniques, and procedures. These efforts took place as part of the "Urban Warrior" experiment, which took place from 1997 to 1999. Urban combat environments were simulated in miniature at Marine bases like Camp Lejeune.[25]

At the same time, a Department of Defense directive established the Marine Corps as the Executive Agent for nonlethal weapons. The Lab was at the forefront of this effort, establishing a directorate specifically for nonlethal weapons.[26] Although nonlethal weapons were never a huge priority in terms of overall Marine Corps budget, the Lab continued to invest in a variety of technologies, such as nonlethal ammunition and sticky foam for incapacitating people. Another nonlethal technology that the Lab studied was the Active Denial System, which used millimeter wave radio frequency to induce a feeling of intense burning without causing physical harm.[27]

As the Marine Corps sought to further develop urban and nonlethal technologies, another change in operations was continuing as the Marines increasingly operated at the boundary between conventional forces and special operations forces. Beginning in the early 1980s, the Marine Corps began to investigate what, if any, role it should play in special operations. This examination took on new importance following the establishment in 1986 and 1987 of the United States Special Operations Command (USSOCOM).

USSOCOM brought together all the disparate elements of special operations from the other three services. Commandant Kelley chose not to include Marine units in USSOCOM, instead choosing to retain all Marine assets in the Corps. He was quoted saying that he "did not believe that committing Marine forces to a fledgling and separate command allowed the corps to retain the level of flexibility it needed to meet the broad spectrum of missions Marine forces were assigned around the world."[28]

However, he did initiate the creation of Marine Expeditionary Units (Special Operations Capable), known as MEU(SOC)s. These units were regular Marine Expeditionary Units (MEUs) that had received additional training to conduct some operations, such as vehicle/ship boarding and seizure and building breaching and assault. Though very capable, the MEU(SOC) could not compare to other special operations forces such as the Army's Delta Force or Navy's SEALs.[29]

Many special operators looked askance at the Marine Corps. Some blamed the 1980 Desert One debacle on the Marine Corps. Others felt that the Marines had an exaggerated sense of their own special operations capability and eliteness. Friction also occurred over the Marine barracks bombing in Lebanon, as some special operators had tried to warn the Marines that they were dangerously vulnerable.[30]

Yet the Marines, particularly the MEU(SOC)s, were often the only substantial force available on short notice to a theater commander. The MEU(SOC)s thus began to operate at a level just above that of the special operations forces alone, but below a major deployment. Noncombatant evacuations represented a substantial portion of these operations. These missions required greater numbers and lift assets than the special operations forces generally had available. But they did not require the level of covertness of most special operations. For example, Marines from the 4th Marine Expeditionary Brigade working with Navy SEALs conducted Operation Eastern Exit, the evacuation of US embassy personnel from Mogadishu, Somalia, in 1991.[31]

By the mid 1990s, the Marine Corps had adjusted bureaucratically to the new strategic environment. The Corps continued to strike a balance between two broad missions, high-intensity ground warfare/amphibious assault and low-intensity conflict/intervention, which were widely accepted both in the organization and amongst policymakers. It claimed unique competences for itself by seizing niche tasks relevant to its broader missions, such as urban warfare and nonlethal weapons. These tasks solidified the Marine Corps' bureaucratic position at relatively low cost to itself, an important achievement in a time of declining military budgets.

The revolution in military affairs and the later interwar years

The second half of the second interwar period was marked by the introduction (or more accurately, elaboration of) the "revolution in military affairs." The RMA concept dated back to the 1970s, when both the Soviet and US militaries began examining the possibility of using information technology to create lightweight "reconnaissance-strike complexes" that would be capable of quickly and efficiently destroying armored vehicles. The first Gulf War showcased the power of precision weapons, and thus provided some empirical support for the RMA. By 1996, both civilian analysts and Air Force and Navy officers argued that the combination of computer/sensor networks and precision weapons had made the RMA a reality.

The Marine Corps was somewhat skeptical toward the RMA, at least in the sense of it being a revolution. Commandant Krulak's writings, for example, stressed the idea that the Marine Corps was about producing skilled Marines rather than relying heavily on technology. A common refrain among Marines was that the other services "man their equipment" while the Marine Corps "equips its men."[32]

Krulak also promulgated such concepts as "the strategic corporal" and the "three-block war," which focused on low-intensity conflict and heavily emphasized personnel over technology.[33] The Marine Corps tended to be equally skeptical of such "Buck Rogers" technology. The attitude of many Marines toward RMA advocates is perhaps best summed up by the cartoon in Figure 6.1, from the Naval Institute's journal, *Proceedings*, from 1999.

At the same time, Krulak and other Marine Corps leaders took pains to show that the Marines were not Luddites. The Marine's most important acquisition programs in this period – the V-22 Osprey and the Advanced Amphibious Assault Vehicle (AAAV) were technologically ambitious. The V-22 Osprey was a tilt-wing rotary aircraft that had been under development for well over a decade by 1996. Despite developmental problems, the Marine Corps continued to fund the V-22 and began to deploy it in 2007.[34] The AAAV was a high-speed replacement for the present Amphibious Assault Vehicle.

Both vehicles were integral to the Marine Corps' major doctrinal shift of this period, termed Operational Maneuver from the Sea (OMFTS). OMFTS is the Marine component of the broader Navy concept "…From the Sea." Promulgated by Commandant Krulak in 1996, OMFTS is an evolution of the maneuver warfare doctrine of the 1980s. It called for the Marines to embrace the RMA as an enabler of maneuver warfare and to be prepared to move rapidly and decisively against enemy centers of gravity. OMFTS thus more closely linked the amphibious part of the Marine Corps to the part that embraced armored warfare.[35]

The tactical concept of Ship-to-Objective Maneuver (STOM) supported OMFTS. STOM sought to bypass the beach as much as possible in conducting

Figure 6.1 The Marine perspective on transformation.

OMFTS. So Marine units that previously had to seize and sustain a beachhead before moving inland would instead move directly to the objective. The V-22, which combined the speed of a fixed-wing aircraft with the vertical take-off and landing of a helicopter, was critical to STOM.[36]

The AAAV was equally important, as it was fast enough to keep up with Marine tanks moving cross-country on land and nearly as fast in the water. It would shorten the time needed to get ashore and could execute the high-tempo operations required by maneuver warfare. Along with Marine aviation assets that provide close air support, AAAV and V-22 would allow amphibious maneuver warfare without a beachhead.

Of course, Marine maneuver warfare still faced the logistics constraints that had earlier led to both land-based and maritime prepositioning. Sea-based logistics in support of OMFTS would be critical for the concept to work. Otherwise, Marine forces might rapidly maneuver ashore only to run out of gas like Patton's Third Army in World War II. The Marine Corps thus had to find a way to rapidly move supplies from what were essentially the successors to maritime prepositioning ships to the shore. Unlike previous Marine operations, there would be no intermediate point on the beach or in a port where supplies could be offloaded and then directed to units.

This rapid movement of supplies would require reliance on information technology and what has been termed "just-in-time logistics" or "precision logistics." Derived in part from both the commercial shipping world and from lean manufacturing techniques pioneered by Japanese automotive firms, precision logistics seeks to transport exactly what a given warfighter needs (no more and no less) to him exactly when he needs it. These efficiency gains would reduce the amount of supplies that need to be transported and thus the effort needed to sustain forces ashore.[37]

The application of just-in-time logistics to warfare has limits. Compared to commercial shipping or automobile manufacturing, combat has inherently higher levels of uncertainty and risk. Any method of logistics would have to be very responsive and operate at high speed for the concept to work in battle.

Another important element of OMFTS was the ability of aviation to provide firepower to ashore units. This aviation would be provided by both the Marines and the Navy. Arguments were made that the two should then integrate more closely. One step down this road took place during the early 1990s, when an initial four Marine F-18 squadrons began to operate as part of Navy carrier air wings. The first to do so was VMFA-314, which embarked on the USS *Abraham Lincoln* in 1992.[38]

Navy–Marine Corps aviation integration would have to be more than just putting Marines on carriers if OMFTS was to work. In addition to adjusting command and control relationships, the Navy would have to absorb the Marine focus on close air support/battlefield air interdiction that was integral to enabling OMFTS. This meant, for example, a Marine liaison to Navy air-to-ground training.

The Marine Corps after September 11

In the wake of the attacks of September 11, 2001, the Marine Corps was called on to execute OMFTS under extremely challenging circumstances. Afghanistan, a land-locked country hundreds of miles from the nearest ocean, was to be the first combat test of OMFTS. Marines from the 15th and 26th MEUs were tasked to establish a forward operating base inside Afghanistan almost as soon as air operations against the Taliban government of Afghanistan began in early October. In fact, Marines from the 15th MEU secretly moved into a Pakistani airbase on October 7, the day the air war began, tasked to provide security for US combat search and rescue forces.[39]

The movement into Afghanistan marked the first time a Marine officer, in this case Brigadier General James Mattis, commanded a Navy task force – Task Force 58, comprised of two Navy amphibious ready groups carrying the two MEUs. This decision followed from that fact that Mattis, commander of the MEB, would be planning and directing the operations ashore.

After establishing the combat search and rescue outpost in Pakistan, the Marines moved a reinforced company into another Pakistani airbase in preparation for going into Afghanistan. Given Pakistan's sensitivity to having US forces on its soil, this action had to be conducted as quietly as possible – making the missions similar to a special operations-type operation, not covert but very discreet.

The application of Ship to Objective Maneuver came on November 25, 2001, when six CH-53 helicopters, accompanied by AH-1W attack helicopters, flew over 400 miles from Task Force 58 to Afghanistan, landing at a desert airstrip near Kandahar. The airstrip had earlier been reconnoitered by SEALs to ensure the Marines would not face a hostile force upon landing. Within an hour, Marine KC-130s landed the reinforced company that had been deployed in Pakistan, along with some of their LAVs. The two units quickly established a perimeter, creating Forward Operations Base Rhino. [40]

This operation was a successful application of STOM, but it pointed out shortcomings in the existing Marine force structure and concept of operations. First, the AH-1W attack helicopters, unlike the CH-53s, could not aerially refuel. They refueled at Shamsi, in Pakistan. This option might not always be available. Similarly, only the ability to pre-stage the LAVs at an airbase in Pakistan made their early deployment in Afghanistan possible. Had Pakistan been less willing to cooperate, the Marines at Rhino might well have had neither gunship support nor armored vehicles.

The V-22 Osprey could help with this shortfall but not eliminate it. For example, a gunship version of Osprey would have to be purchased to mitigate the attack helicopter problem. The Osprey would also not have solved the armored vehicle transport issue. It cannot carry the load.

Another issue was logistics. The Marines did not face severe shortages, but the base was very austere. General Mattis chose to build up a seven-day stockpile and to also bring forward additional medical resources. This was undoubtedly prudent, but points to the possible limitations of "precision logistics." While logistics in this case certainly were more efficient than the fabled "iron mountain" of supplies the Marines established for the 1991 Gulf War, they were certainly not revolutionary.

Even as Marines established Forward Operating Base Rhino, planning was beginning for the invasion of Iraq. The Marine Corps would have a major part in this operation. The eastern part of the push to Baghdad would be undertaken by the Marines' I Marine Expeditionary Force, under Lt. Gen. James Conway. The major elements of this operation were the 2nd Marine Expeditionary Brigade and the 1st Marine Division. This division was led by the recently promoted Major General Mattis, but it looked quite different from the operation he commanded in Afghanistan. Rather than STOM, the Marine drive to Baghdad was the longest Marine armored movement in history. It relied on the normal logistics chain, and its M1 tanks expended huge amounts of ordnance and burned large quantities of fuel.[41]

There were some new aspects to the operation, to be sure. Like the Army's corresponding movement to Baghdad, this drive relied on many of the networks and systems developed and fielded during the previous decade. Mattis also emphasized lean logistics. He sought to use captured fuel and told Marines not to waste their food, as they might get only one MRE a day. Mattis also sought to capitalize to the maximum extent possible on aviation to provide firepower – a tactic allowed by precision airpower.

Yet, overall there was nothing about the effort that could be called revolutionary. It was conducted the way Marines had been planning to operate for at least two decades.[42]

By February 2004, with the insurgency well underway, I Marine Expeditionary Force redeployed to al-Anbar province, the heart of the Sunni insurgency. The violence in Anbar began to worsen rapidly, culminating in the killing of four private security contractors in the city of Fallujah in April. The Marines found themselves embroiled in almost exactly the kind of conflict that Commandants Gray and Krulak had been arguing the Corps was likely to face. Cities like Fallujah and Ramadi had become insurgent-dominated strongholds. Telling friend from foe was difficult. In Fallujah, after one abortive attempt to retake the city, this problem was solved by warning the inhabitants to leave and then attacking, using some of the techniques and equipment developed by the urban warfare studies of the 1990s.[43]

Even as they conducted a major operation against Fallujah, the Marines were writing new doctrine for small wars. This included the joint manual FM 3–24/ MCWP3–33.5 *Counterinsurgency Operations*, as well as the *Small Unit Leader's Guide to Counterinsurgency*. The Marine Corps, which had spent the better part of the second interwar period talking about low-intensity conflict but preparing for high-intensity conflict and rapid maneuver, was forced to begin rethinking its approach to combat. Yet this, too, was less a revolution than a return to the Marine emphasis on small wars that had been eclipsed by the maneuver warfare advocates after Vietnam.

Maneuver warfare did not disappear with the resurgence of low-intensity conflict. The term "expeditionary" as defining a unique Marine Corps capability became increasingly common during this period, so "maneuver warfare" became "expeditionary maneuver warfare" and the AAAV became the Expeditionary Fighting Vehicle. The Marine acquisition of the Expeditionary Fighting Vehicle and Osprey continued.[44]

An additional term has also become prominent in Marine doctrine since the end of the second interwar period. This is the concept of "distributed operations," which has the flavor of network-centric operations (small dispersed units connected via networks) but also implies that considerable autonomy will be given to small-unit (battalion and below) commanders. At the same time, distributed operations is said to apply equally to counter-insurgency and high-intensity warfare. It seems to be an attempt to find a unifying element of doctrine that is common to both low-intensity conflict and maneuver warfare advocates.[45]

On the technical front, Marines have made nearly continuous use of UAVs in Iraq and Afghanistan, ranging from large Air Force UAVs like Global Hawk and Predator (when available) to smaller assets like Dragon Eye. As yet, these UAVs have not become sufficiently ubiquitous to constitute a revolution. That is, even if UAVs were far more common, they would remain an adjunct or addition to current ways of operating.[46]

In addition to operations on the ground, other changes were taking place in the Marine Corps. In 2004, Navy Secretary Gordon England pushed forward the initial, tentative integration of Navy and Marine Corps aviation. The proposal was to have six more Marine squadrons assigned to carrier air wings at any given time and three Navy squadrons supporting Marine units. This would reduce the total needed procurement of such items as the F-35 Joint Strike Fighter, as well as force Navy fighter squadrons to maintain a higher overall level of readiness. (Normally Navy squadrons suffer very low readiness in the post-deployment period.)

The Memorandum of Agreement signed in November of 2005 between the Navy and Marine Corps paints an ambitious transformational vision:

> The Chief of Naval Operations' (CNO) and Commandant of the Marine Corps' (CMC) transformation vision outlined in Navy Power 21 merges Navy and Marine Corps TACAIR into a seamless Naval Aviation force at sea and ashore.... The goal is to exploit revolutionary network-centric warfare and expeditionary maneuver warfare concepts to enhance power projections by tightly integrating Carrier Strike Groups (CSGs), Expeditionary Strike Groups (ESGs), and Marine Air-Ground Task Forces (MAGTFs).[47]

It is unclear how much of this plan will occur. There are several threats to Marine Corps interests in this program. According to some in the Marine Corps, the Corps is not overjoyed by the plan, though they are not resisting. Many Marine versions of the Joint Strike Fighter were to be a short takeoff/vertical landing version that could operate from austere airfields. However, this version is not well-suited to carrier operations, mostly because of its more limited range. Moreover, Marines fear that this is the end of their organic aviation capability. Finally, if Navy pilots prove less willing and skillful in providing close air support than Marine pilots, the Corps may revolt against the integration.

The final change that has happened in the Marine Corps in recent years is a major reversal on its position toward US Special Operations Command. After Secretary of Defense Donald Rumsfeld gave USSOCOM the lead in the "Global War on Terror," he essentially directed the Marine Corps to start providing forces to USSOCOM.[48] In response, the Marine Corps created its own special operations component, Marine Special Operations Command (MARSOC), which is now part of USSOCOM. MARSOC itself has two main elements. The Marine Special Operations Companies (MSOCs) are the direct descendants of Marine Force Reconnaissance Companies with additional training. However, for now, they are not independently deployable like Navy SEAL teams. Instead they remained tied to a parent MEU, as the Force Reconnaissance companies were previously. The other elements of MARSOC are Foreign Military Training Units, which are twelve-man teams that seem to be modeled on Army Special Forces A-teams. They have the similar mission of training foreign militaries and possibly conducting unconventional warfare.[49]

MARSOC raises a number of bureaucratic questions for the Marine Corps. The Corps has always prided itself on its elite esprit de corps. MARSOC will be an elite within an elite, which could cause tension. As a result, MARSOC may not last. If it does survive, MARSOC will raise questions about officer selection and promotion. All the other services more or less segregate the special operations officer corps from the "main line" of officers. The Marines may be unable or unwilling to do this given their size, which in turn may mean MARSOC looks little different than regular Marines. Or it could mean that MARSOC Marines come to dominate the service. Neither would be optimal. Finally, it is unclear whether MARSOC will stay close to the Marines or drift into the orbit of USSOCOM, as other special operations forces have.[50]

Conclusion

The Marine Corps' experience since the Cold War is an example of evolutionary rather than revolutionary change. Where helpful, technology was welded onto to existing operational concepts and force structure. UAVs and networks linking ground units and aviators improved the Marines Corps to defeat opposing units, but did not overturn their methods of doing so. Advanced Amphibious Assault Vehicles and V-22s brought more change – theoretically allowing Marines to attack coastlines without fighting for a beachhead – but this is more a method of preserving the amphibious assault mission than revolutionary doctrine. Precision weapons allowed the Marines to rely less on artillery, but have not replaced it. The underlying concepts and structures of the Corps have changed very little since the Cold War – no conventional transformation occurred.

The central changes in the Marine Corps during this period had more to do with the ascendance of an intra-organizational community (maneuver warfare advocates) than with any kind of technological change. But this change is a shift toward an old Marine Corps competence, not a revolution. Other changes of interest – the development of UAVs and nonlethal technologies, the emphasis on MOUT, and the development of MARSOC – can be traced at least in part to inter-service pressures. In order to protect its primary missions of interest, the Marine Corps was willing to make minor doctrinal and force changes for reasons of political salesmanship.

Neither did environmental transformation occur during the second interwar period. In large part this can be attributed to the vicissitudes of Rosenian intra-service politics. Small wars advocates like Commandants Gray and Krulak could make minor changes when running the Corps, only to have the organization go in a different direction after they left their positions. More generally, the dominance of the maneuver warfare community in the Corps was an impediment to transformation of all kinds. This community had no interest in changing the way it did business, which creates a tough environment for potential innovators.

The wars in Afghanistan and Iraq have returned small wars to the fore, and could have major ramifications for force structure and operations. The rise of counter-insurgency advocates, the creation of MARSOC, and the progression of Navy–Marine Corps aviation integration could all lead to force structure changes that would be worthy of being called transformation, if not revolution. However, it seems more likely that the Marine Corps will continue its traditional course of hedging between its primary missions and taking on niche missions that complement these missions – assuring its relevance by retaining expertise in various realms of warfare.

Notes

1 See Keith B. Bickel, *Mars Learning: The Marine Corps' Development of Small Wars Doctrine, 1915–1940* (Boulder, CO: Westview Press, 2001); and Kenneth J. Clifford, *Amphibious Warfare Development in Britain and America from 1920–1940* (Laurens, NY: Edgewood, 1983).

2 See *Title 10 of the US Code*, section 5063. On Marine insecurity, see Victor Krulak, *First to Fight: An Inside View of the US Marine Corps* (Annapolis, MD: Naval Institute Press, 1984), pp. 15–66.

3 For a recent example, see Jim Garramore, "Fallujah Battle Highlights Marine, Joint Capabilities," *American Forces Press Service*, February 5, 2005. Online, available at: www.defenselink.mil/news/newsarticle.aspx?id=25912 (accessed September 14, 2007.)

4 "CAPEX Held at Camp Lejeune," *Marine Corps Gazette*, Vol. 83, No. 9 (September 1999).

5 On the northern flank see, for example, William K. Sullivan, "Soviet Strategy and NATO's Northern Flank," *Army War College Strategic Studies Institute Paper* (Carlisle, PA: US Army War College, 1978); Marian K. Leighton, *The Strategic Threat to NATO's Northern Flank* (New York, NY: National Strategy Information Center, 1979); Jeffrey G. Barlow, "NATO's Northern Flank: The Growing Soviet Threat," *Heritage Foundation Backgrounder*, No. 82 (May 1979). For a skeptical view, see Barry R. Posen, "Inadvertent Nuclear War? Escalation and NATO's Northern Flank," *International Security*, Vol. 7, No. 2 (Fall 1982), pp. 28–54.

6 Sean M. Maloney, "Fire Brigade or Tocsin? NATO's ACE Mobile Force, Flexible Response, and the Cold War," *Journal of Strategic Studies*, Vol. 27, No. 4 (December 2004), pp. 606–608.

7 For discussion of the planning and decision on where to locate the equipment, see Michael Vrabel, "Can We Move the Men and Equipment to North Norway in Time?" Marine Corps Command and Staff College Paper (1988).

8 See http://hqinet001.hqmc.usmc.mil/pp&o/POE/POE-60/mcppn.asp (last accessed April 23, 2007) for a history of the NALMEB program, which was subsequently renamed the Marine Corps Prepositioning Program Norway. For discussion of the broader argument for the Marine Corps involvement in NATO, see Linton Brooks, "Naval Power and National Security: The Case for the Maritime Strategy," *International Security*, Vol. 11, No. 2 (Autumn 1986).

9 "A Discussion of the Rapid Deployment Force with Lieutenant General P. X. Kelley" (Washington, DC: American Enterprise Institute, 1980).

10 Bernard Burchell, "Maritime Prepositioning Force: A Historical Analysis," Marine Corps Command and Staff College Paper (1993).

11 Karl Gunzelman, "White Knight or White Elephant: The M1A1 'Abrams' in the Marine Corps," Marine Corps Command and Staff College Paper (1989) and John Sykes, "The Marine Corps Must Have Tanks," Marine Corps Command and Staff College Paper (1989).

12 Joseph Crookston, "Marine Corps Roles and Missions: A Case for Specialization," Marine Corps Command and Staff College Paper (1987).

13 See Marine Corps FMFM 1, *Warfighting* (1989) and P. B. Retter, "Embracing Maneuver Warfare and for War," Marine Corps Command and Staff College Paper (1989).
14 Dennis Beal, "The M1A1 Tank: Its Role in the Marine Corps," Marine Corps Command and Staff College Paper (1991).
15 T. X. Hammes, "Insurgency: The Forgotten Threat," *Marine Corps Gazette*, Vol. 72, No. 3 (March 1988) and Eric Nyberg, "Insurgency: The Unsolved Mystery," Marine Corps Command and Staff College Paper, 1991.
16 Molly Moore, "Marine Commandant Striving for Leaner, Meaner Fighting Machine," *Los Angeles Times*, February 5, 1989.
17 See *Jane's Armour and Artillery* database entry for M1 Abram (March 2007) for a discussion of Marine acquisition of the M1. Online, available at: http://jaa.janes.com/.
18 Thomas Ehrhard, *Unmanned Aerial Vehicles in the United States Armed Services: A Comparative Study of Weapon System Innovation* (PhD thesis, Washington, DC: The Johns Hopkins University, June 2000).
19 Mark Bowden, *Black Hawk Down: A Story of Modern War* (New York, NY: Atlantic Monthly Press, 1999).
20 See discussion in Kevin Brown, "The Urban Warfare Dilemma – US Casualties vs. Collateral Damage," *Marine Corps Gazette*, Vol. 81, No. 1 (January 1997).
21 See discussion in Frederick Lorenz, "Less-Lethal Force in Operation United Shield," *Marine Corps Gazette*, Vol. 79, No. 9 (September 1995).
22 David Patton, "Operation Uphold Democracy: The Use of Military Police and Nonlethal Weapons in Haiti," RAND MOUT Conference Proceedings, online, available at: www.rand.org/pubs/conf_proceedings/CF148/CF148.appo.pdf (last accessed April 23, 2007).
23 "The Commandant's Warfighting Laboratory," *Marine Corps Gazette*, Vol. 79, No. 9 (September 1995).
24 Interview with Charles Krulak, October 22, 2000. Online, available at: www.worldnetdaily.com/news/article.asp?ARTICLE_ID=19659 (last accessed April 23, 2007).
25 Gary Schenkel, "The Transition from HUNTER to URBAN WARRIOR," *Marine Corps Gazette*, Vol. 81, No. 12 (December 1997) and "URBAN WARRIOR: First Impressions," *Marine Corps Gazette*, Vol. 83, No. 6 (June 1999).
26 John Kelly, "Simulation Support for Nonlethal Weapons," *Marine Corps Gazette*, Vol. 81, No. 8 (August 1997).
27 See Active Denial factsheet, online, available at: www.jnlwp.com/Resources/FactSheets/ADSFactSheetFinalMarch2007.pdf (last accessed April 23, 2007).
28 Quoted in Mark Clark, "Should the Marine Corps Expand Its Role in Special Operations?" Army War College Paper (2003).
29 Ibid.
30 Michael Smith, *The Killer Elite: The Inside Story of America's Most Secret Special Operations Team* (London: Cassel, 2006), Chapters 1 and 4.
31 See www.navy.mil/navydata/policy/vision/vis00/v00-a.html (last accessed April 23, 2007).
32 See, for example, Charles Krulak, "Commandant's Planning Guidance Frag Order," *Marine Corps Gazette*, Vol. 81, No. 10 (October 1997).
33 Charles Krulak, "The Strategic Corporal: Leadership in the Three-Block War," *Marine Corps Gazette*, Vol. 83, No. 1 (January 1999).
34 Dave Montgomery, "First Squadron of V-22s Quietly Deployed to Iraq," *Fort Worth Star Telegram*, September 19, 2007.
35 Charles Krulak, "Operational Maneuver from the Sea," *Joint Forces Quarterly*, Vol. 80, No. 6 (Spring 1999), pp. 78–86.
36 Michael Peck, "Marines Sketch 'Expeditionary Maneuver Warfare' Scenarios," *National Defense* (October 2003).

37 Ronald Fricker and Marc Robbins, *Retooling for the Logistics Revolution: Designing Marine Corps Inventories to Support the Warfighter* (Santa Monica, CA: RAND, 2000).
38 See www.3maw.usmc.mil/mag11/vmfa314/history.asp (last accessed April 23, 2007).
39 Arthur Brill, "Afghanistan Diary," *Seapower* (April 2002).
40 Brill, "Afghanistan Diary."
41 Bernard Trainor and Michael Gordon, *Cobra II: The Inside Story of the Invasion and Occupation of Iraq* (New York, NY: Vintage Books, 2005), pp. 182–190.
42 Bing West, *The March Up: Taking Baghdad with the United States Marines* (New York, NY: Bantam, 2004).
43 Bing West, *No True Glory: A Frontline Account of the Battle of Fallujah* (New York, NY: Bantam, 2006).
44 Congressional Budget Office, *The Long-Term Implications of Current Defense Plans* (Washington, DC: Congressional Budget Office, January 2007).
45 Vincent Goulding, "Distributed Operations," *Marine Corps Gazette*, Vol. 91, No. 2 (February 2007).
46 Alfred Connable, Christopher Olson, Gregory Smietanski, and Christopher D. Kondaki, "Better Eyes for the Infantry," *Marine Corps Gazette*, Vol. 89, No. 12 (December 2005).
47 "Department of the Navy Tactical Aircraft Integration," *Memorandum of Agreement Between Commander, Naval Air Forces, and Deputy Commandant for Aviation*, United States Marine Corps, November 23, 2005, p. 1.
48 Jack Glasgow, "A Time for Every Purpose," *Marine Corps Gazette*, Vol. 90, No. 1 (January 2006).
49 Dennis J. Hejlik, Cliff W. Gilmore, and Matthew P. Ingram, "Special Operations Marines and the Road Ahead," *Marine Corps Gazette*, Vol. 90, No. 8 (August 2006).
50 See Chapter 7 for a discussion of how special operations units associated with particular services are more properly thought of as a part of a new Special Operations service.

7 The fifth service

The rise of Special Operations Command

Colin Jackson and Austin Long

For most of the Cold War, Special Operations Forces (SOF) occupied an odd niche in the US military. Marginalized by their respective services, they survived by offering civilian policymakers capabilities that the core services did not and often disparaged. Dependent on the imperfect and intermittent protection of civilians, and lacking a unified organizational structure or budget authority, SOF survived by maintaining a very broad portfolio of undesirable missions.

No community underwent greater transformation during the second interwar period than did Special Operations. Several major changes propelled Special Operations Command (SOCOM) from a marginal role to the status of a "fifth service." The rising power and autonomy of SOCOM led it to reprioritize its mission areas. While increased funding has expanded opportunities for nearly all groups in the Special Operations community, SOCOM has invested more heavily in direct action (DA) than in unconventional warfare (UW). The global war on terrorism (GWOT) heightened SOCOM's autonomy and capability, but these changes, paradoxically, led it to discard many of the missions that policy-makers need most in combating terrorism. A strong internal preference for direct action, amplified by early civilian enthusiasm for direct action strategies, skewed the American approach to counterterrorism, leaving national decision-makers long on direct action assets and short on unconventional warfare options.

This chapter proceeds in five parts. The first section briefly describes the various US SOF units and their organizational histories. The second presents the rise of joint SOF organizations from 1979 to the end of the Cold War. The third section discusses the experience of SOF during the second interwar period. The fourth section describes the challenges and successes of SOF after September 11. The final section analyzes the problems and prospects for this fifth service and their implications for transformation.

Missions and preferences

Throughout the Cold War, the special operations community lived in the shadow of the four services. Viewed with suspicion and contempt by their parent services, special operations forces were forced to make do with limited funding. Their subordinate status within each service led most community leaders to pursue a basket of missions. Though the community might have preferred to pursue only some subset of these missions, prudence and bureaucratic vulnerability made a diversified mission portfolio desirable. As a result, SOF took on at least nine primary missions: direct action (DA), unconventional warfare (UW), special reconnaissance, foreign internal defense (FID), counterterrorism, counter-proliferation (CPT), civil affairs (CA), psychological operations, and information operations (IO). Although each of these missions was distinct, special operators often divided these into two general categories: direct action (to include direct action, special reconnaissance, counterterrorism, and counter-proliferation) and unconventional warfare (unconventional warfare, foreign internal defense, civil affairs, psychological operations, and information operations). As the term suggests, direct action involved the independent action of special operators to conduct raids or reconnaissance missions; unconventional warfare missions are executed "by, with, and through" the use of local forces.[1]

Three major changes in the late Cold War and interwar periods greatly enhanced the status of special operations forces. The first was organizational. The Defense Authorization Act of 1987 established the Special Operations Command (SOCOM): a four-star unified command with responsibility for all special operations activities.[2] Though this reorganization stopped short of making SOF a true fifth service, it did bring all special operations forces under a single, major command. The creation of the Major Force Program 11 (MFP-11) budget program in 1989, which is discussed below, put SOCOM on the road to limited budget autonomy.[3] These civilian decisions to place special operations under a separate command and extend them substantial funding autonomy led to a steep change in the latitude and clout of SOCOM.

The second change was technological. Historically, SOF had played important niche roles but had lacked the combat power to influence outcomes on the battlefield. The arrival of three RMA technologies in particular – precision-guided munitions (PGMs), global positioning systems (GPS), and satellite communications – greatly increased the power and compatibility of SOF on the conventional battlefield. The prominent role played by PGMs in Desert Storm led to a major expansion in the number and types of weapons developed by the services. Growth in PGMs put a premium on sensors that could provide precise targeting data, and special operators were a natural fit for this role. Equipped with laser designators, SOF personnel could give airborne forces the precise targeting data necessary to deliver very effective aerial fires. More important, operators could fulfill this targeting role with only a fraction of the number of personnel present in conventional ground units. To advocates of radical transformation, the combination of air-delivered PGMs and special operations ground targeting offered a way to move beyond the traditional model of land combat.[4]

The same technological changes that gave SOF greater clout on the battlefield also brought them closer to the military mainstream. Advanced communications systems made it far easier for geographically dispersed units to communicate with aircraft, adjacent ground units, and theater and national headquarters. Long-range penetration missions, once considered "fire and forget" missions, could be monitored and controlled from great distances.[5]

The third major change was the arrival of the war on terror. Although SOF played a more important role in various small wars of the period, especially Vietnam, the conventional nature of the Cold War standoff in Western Europe had virtually guaranteed a subordinate role for special operations forces.[6] The war on terror, by contrast, was an irregular war of unlimited duration, a campaign more suited to SOF's strengths. Secretary Rumsfeld's decision to designate SOCOM as the lead or "supported" command for the GWOT made the centrality of SOF concrete.[7] The unlimited duration and vague definition of the threat made SOCOM's status elevation effectively permanent.

SOCOM's new power and status led to a major shift in its internal priorities. A more vulnerable SOF community had been willing to perform a range of roles to safeguard its relevance in a competitive budget environment. A more independent and muscular SOCOM felt confident enough to focus its efforts on its preferred mission set – direct action.[8] Long the province of secret, special mission units or "black" SOF, DA became a core mission for "white" or non-secret units as well. Starting in the major combat phase of Afghanistan and continuing through the two counterinsurgency campaigns in Iraq and Afghanistan, white SOF demonstrated the ability to perform direct action and counterterrorism missions on a rough par with black SOF.

Intraservice competition between white and black SOF for the high prestige counterterrorism missions has created a new set of problems for policymakers and military commanders. The rush to direct action has left other missions, particularly unconventional warfare, foreign internal defense, civil affairs, and psychological operations, in decline. The effectiveness of direct action in counterterrorism or counterinsurgency rests on an implicit theory of victory. By killing or capturing a list of high value targets, operators assume that over time they will weaken or defeat their opponents.[9] This theory, a close relative of air force theories of decapitation, shares little in common with the unconventional warfare (UW) and foreign internal defense (FID) theories of victory.[10]

In unconventional warfare and foreign internal defense missions, the role of the operator is to build the capability of indigenous forces to perform military and paramilitary functions. In some instances, the object is to use these indigenous forces to accomplish the desired objectives. In others, unconventional warfare is a means to garner goodwill and with it the information necessary to conduct future direct action operations. In still other cases, the training and inclusion of local forces may simply be a means to assuage bureaucratic or political sensitivities surrounding the activities of American special operators in the host country. In these cases, indigenous forces accompany American special operators simply to provide a political fig leaf for American special operators to perform their own direct action missions.

What the technological, bureaucratic, and mission shocks of the past three decades have revealed is the internal preference ordering and status hierarchy of the SOF community. Given the choice between the uncertain and long-term investments involved in unconventional warfare, and the certain and immediate rewards of a successful direct action operation, most operators would choose the latter.[11] The problem is that most irregular warfare, and particularly counterinsurgency, relies on a balanced mix of these operations. Without SOF involvement in the training and leadership of indigenous forces, policymakers are forced to rely on conventional forces to train local forces. Though some conventional forces succeed in this mission, they generally lack the training and language facility to match the support provided by SOF. When the training of indigenous forces suffers as a result of these priorities, the occupying force is often unable to extract itself from an endless cycle of successful but strategically indecisive direct action. Ironically, though SOCOM received the lead in the war on terror in recognition of its skills and history in both direct action and unconventional warfare, the new authority, autonomy, and resources extended to SOCOM have led the organization away from unconventional warfare and toward a gratifying but largely untested set of theories about critical node targeting.

Who's who in the zoo: SOF units

SOF, by its nature often clandestine, comes in two basic types: "white" and "black." White SOF are units whose existence is openly acknowledged by the US government even if their operations are almost always classified and clandestine (clandestine meaning that the effect is observable even if it cannot be attributed). White SOF consists of the US Army Special Forces, psychological operations, civil affairs, the 75th Ranger Regiment, and the 160th Special Operations Aviation Regiment (SOAR); the US Navy's SEALs; the US Air Force's various SOF aviation units and combat controllers, pararescue, and weather support; and, more recently, the US Marine Corps' Foreign Military Training Units and Special Operations Companies.

Black SOF are units (often referred to as special mission units or SMUs) whose existence is not acknowledged and whose operations are not only always classified/clandestine, but often covert, meaning that the operation is not readily observable and can be plausibly denied by the US government. Black SOF is thus more akin to covert intelligence operatives, such as the Central Intelligence Agency's Special Activities Division (SAD), and black SOF and SAD work together frequently. Those SMUs whose existence is known but not acknowledged are the US Army's Delta Force and Intelligence Support Activity and the US Navy's SEAL Team 6.[12] Below is an overview of both white and black SOF units. This summary is derived from unclassified sources and is therefore tentative.

The Army Special Forces (SF) were born in the first two decades of the Cold War.[13] Many Eastern Europeans in the United States joined with veterans of US clandestine efforts during World War II (or were already veterans of those efforts) to form the nucleus of SF. The 10th Special Forces Group (Airborne) deployed to Germany in 1953 and focused on Europe. Subsequently other groups were formed with different geographic orientations: 1st Group (Asia), 3rd Group (Africa); 5th

Group (the Middle East); 7th Group (Latin America). SF operators generally seek to acquire a broad range of noncombat skills, especially languages.[14]

Along with SF, civil affairs (CA) and psychological operations (PSYOP) units are also part of the Army's umbrella SOF command, known as United States Army Special Operations Command (USASOC). CA units focus on liaising with local populations and providing infrastructure development and support during combat operations. PSYOP units seek to provide targeted information operations in support of US objectives, ranging from leaflets to radio broadcasts. Both often work closely with SF.

The 75th Ranger Regiment is a descendant of several different scouting/raiding organizations dating back to before American independence, most notably Roger's Rangers in the French and Indian War. The US Army created various Ranger units during World War II, Korea, and Vietnam. These elite units were disbanded after the wars ended. However, after Vietnam, concern about the need for rapidly deployable light infantry led Army Chief of Staff Creighton Abrams to establish a peacetime Ranger organization that would eventually reach regimental size. The Rangers specialize in forced entry operations, including the seizure of airfields and the provision of security for black SOF DA operations.

The 160th SOAR was created in the aftermath of the failed Iranian hostage rescue of 1980, which failed largely due to aviation-related problems. The 160th specializes in night flying in both specialized transport helicopters and in attack/scouting helicopters.

The US Navy's SEALs (an acronym for Sea, Air, and Land) grew from combat diver and underwater demolition teams from World War II and Korea. The SEALs were formally established in 1962 at the behest of President Kennedy to provide a maritime/riverine counterpart to many of the Army's SF activities. SEALs possess an array of combat and survival skills, including combat swimming/diving and reconnaissance. In addition to the SEALs, Navy Special Warfare Command (NAVSPECWARCOM) includes Special Boat Units that support SEALs and other SOF operators.

The US Air Force's Special Operations Command (AFSOC) operates a wide array of aircraft. These include specialized fixed-wing and helicopter transport, fixed-wing gunships, unmanned aerial vehicles, specialized refueling aircraft, and information operations/psychological warfare aircraft. In addition, AFSOC has three main types of operational ground personnel. Combat controllers are forward-deployed air traffic controllers who coordinate air support for SOF operators. Pararescuemen (known as "parajumpers" or PJs) specialize in combat search and rescue (CSAR) in hostile environments. Special operations weathermen forward-deploy to provide precise and timely meteorology information for SOF operators in hostile environments.

The Marine Special Operations Command (MARSOC), activated in February 2006, has two types of units. The Foreign Military Training Units train local forces for use in internal defense or offensive operations. The Marine Special Operations Companies (MSOCs) provide direct action and reconnaissance capability. Since

MARSOC is still in its infancy, the majority of personnel and small units have been drawn from existing Marine Force Reconnaissance or line units.[15]

The Army's Delta Force (also referred to as the 1st Special Forces Operational Detachment-Delta, the Combat Application Group, or simply Delta) is the US ground component SMU. Specializing in DA missions such as capturing buildings, Delta was created as a counterterrorism/hostage rescue unit in the 1970s. It has grown and branched out, performing a variety of DA and special reconnaissance (SR) missions. In joint operations, it is often referred to as Task Force Green or Green.

The Army's Intelligence Support Activity (ISA) is the most covert of the three known major SMUs. It changes cover names on what seems to be at least a biannual basis, and maybe as frequently as every major operation. Though still referred to by many as ISA or simply "the Activity," its alleged code names include Central Spike, Grey Fox, and Torn Victor. It was established after the failed Iranian hostage rescue mission to provide tactical intelligence, both human and electronic, for special operations. It also is alleged to have absorbed a covert aviation unit known alternately as Sea Spray, Flight Concepts Division, and Quasar Talent. ISA has been nicknamed "the Secret Army of Northern Virginia," and is often referred to in joint operations as Task Force Orange or Orange.[16]

SEAL Team 6 (also known as Naval Special Warfare Development Group or DEVGRU) is the Navy's equivalent to Delta. Like Delta, SEAL Team 6 specializes in DA but with particular emphasis on the maritime environment. Like ISA, it was established in the wake of the Iranian hostage rescue attempt. In joint operations, it is often referred to as Task Force Blue or Blue.

There is also a less well-known but publicly acknowledged (i.e. not black) Air Force SMU, the 25th Information Operations Squadron. The 25th provides additional intelligence to SOF and conventional forces. Finally, the Marine Corps has participated in black SOF operations but not on a large scale, providing either individuals (for example, detailing them to ISA) or small reconnaissance/intelligence elements. With the creation of MARSOC, this may change.

Although all four services have contributed to SOCOM, the contributions are far from equal. Army SF has typically dominated the white arena, with the Navy contributing considerable white DA and black SOF units (see Figure 7.1).

Two communities, two mission sets

The black and white SOF communities are separated not simply by their secrecy but also by their traditional roles and historical roots. Traditionally, white SOF has focused on unconventional warfare and foreign internal defense.[17] While the distinctions between these missions can blur at the margins, both involve the use of SOF to train and develop indigenous forces for offensive or defensive activity. Unconventional warfare typically involves the development of guerilla and paramilitary capabilities. To secure the trust of those they train, white SOF units frequently employ a mix of civil affairs and psychological operations. By providing tangible economic benefits to the local communities in which they

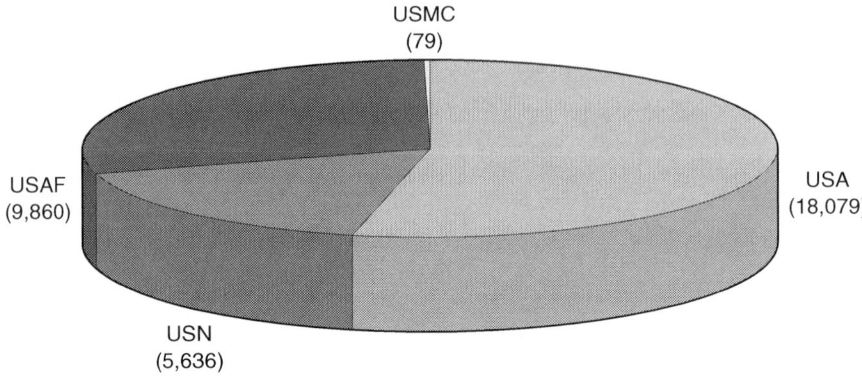

Figure 7.1 Service contributions to SOCOM.

operate, SOF operators hope to foster loyalty and garner actionable intelligence on their opponents. Most unconventional warfare missions are long-duration affairs in which success or failure is only evident after many months of intensive training. The formative experiences of the white SOF community were Vietnam and El Salvador, where the special forces took the lead in training local forces – often clashing with conventional Army forces, which trained local forces to mirror its own offensive tactics rather than the more defensive tactics special operators preferred.

By contrast, black SOF has tended to focus on direct action – the use of commandos to conduct raids or special reconnaissance. Direct action missions involve few of the entanglements or delays of UW. In DA, a handful of operators can execute the task and receive immediate feedback on its success or failure. The historical foundation of the black SOF community is Operation Eagle Claw – the failed mission to rescue the hostages from Iran. That mission ended in disaster, as multiple helicopters broke down at the desert site in Iran chosen for refueling (known as Desert One), and a helicopter and fixed-wing aircraft collided and the helicopter crashed and burned. A second mission was planned (Operation Honey Badger, the aircraft modification part of which was known as Operation Credible Sport) but was not executed.[18]

The debacle at Desert One was investigated by a military commission chaired by former Chief of Naval Operations Admiral James Holloway. The commission's key recommendation was the need for a standing task force focused on counterterrorism and special operations. This recommendation was almost immediately implemented with the establishment of the Joint Special Operations Command (JSOC). JSOC, commanded by a major general, would oversee the SMUs and command black SOF activities. In addition to the SMUs, JSOC often utilizes the 75th Ranger Regiment (known as Task Force Red or simply Red in JSOC operations) and the 160th SOAR.

JSOC operators spent most of the 1980s involved in an array of highly covert activities. Among these were operations to potentially free various hostages, including Brig. Gen. James Dozier, who was kidnapped in Italy by the Red Brigades terrorist organization. JSOC also operated in Honduras and El Salvador in support of US activities in those countries as well as in support of the Nicaraguan rebels known as the contras.[19]

In 1986, Congress passed a Defense Reform Act that is universally known as Goldwater-Nichols after its cosponsors. Among other things, Goldwater-Nichols mandated the creation of joint commands such as the regional Central Command (CENTCOM) and the functional Space Command (SPACECOM). Almost simultaneously, Senators Sam Nunn and William Cohen (among others) were working to increase both the institutional integration and civilian oversight of SOF. They felt SOF needed its own unified command, and they created an amendment to the 1987 Defense Authorization Act that created United States Special Operations Command (USSOCOM or simply SOCOM).[20] SOCOM was under the command of a four-star general, giving it significant organizational clout. The amendment also created a civilian oversight position in the form of the Assistant Secretary of Defense for Special Operations and Low-Intensity Conflict (ASD/SOLIC or simply SOLIC).

With the creation of SOCOM, JSOC became essentially a standing operational element of the new command. Further, JSOC operators quickly came to dominate the leadership of SOCOM. Of the seven non-interim commanders of SOCOM, for example, all but two have been either former commanders or deputy commanders of JSOC (the exceptions are Generals James Lindsay and Hugh Shelton, both of whom had Army SF experience).

Even more significant than the establishment of SOCOM as a four-star command was the creation in 1989 of Major Force Program 11 (MFP-11). While JSOC always had a certain amount of "black budget" money, these sums were never large. Consequently, special operations forces were always highly dependent on their parent services. This limited the ability of SOFs to actually be "special," as they often had to rely on regular service equipment.

Along with provisions in Title X (section 167) of the US code, MFP-11 altered the fiscal balance of power between SOF and the services. This new budget category ensured that SOCOM would have an acquisition budget of its own, a major step toward bureaucratic independence for SOF. However, MFP-11 is limited to acquiring items that are SOF-unique; the individual services must provide common equipment. This restriction has caused problems, as there is often disagreement about what is SOF-unique and what is not. Still, MFP-11 was a major boon to SOCOM and its components, as it gave SOCOM quasi-Title X authority.[21]

Though MFP-11 is small in the context of the overall defense budget, it includes funds for SOCOM operations, maintenance, and procurement, as well as research and development. The impact of MFP-11 is magnified by the small size of the SOCOM force; SOCOM includes 10,000 combat personnel and 40,000 support personnel.[22] Along with the fact that SOCOM does not pay the personnel costs for many of its members, the relatively small SOCOM budget gives it great organizational leverage and autonomy (see Figure 7.2).

Several other factors enhance SOCOM's autonomy. First, the nature of special operations makes SOCOM inherently secretive. Operational secrecy means that SOCOM can shroud many of its operations and bureaucratic activities from the services and the public. Second, it has significant authority to use nonstandard procurement methods (such as no-bid contracts) and to conduct nonstandard operations that increase autonomy.[23]

The most important factor enhancing SOCOM autonomy is the fact that it does not have strong and independent civilian oversight. The ASD/SOLIC oversees a broad range of issues but has only the staff and clout of an assistant secretary, hardly the level of resources and standing available to a four-star general. Further, many ASD/SOLICs are either veterans of SOCOM (especially JSOC) or other covert agencies.

Although having professionals with knowledge and experience in key positions is understandably important, it contributes to opacity and insularity at SOCOM. To paraphrase Max Weber, secrecy is the fighting posture of the bureaucracy.[24] Few organizations can match SOCOM (particularly JSOC) in this regard, given its lack of oversight. Further, this insularity means that oversight may be affected by existing personal relationships.

The long war begins

The attacks of September 11, 2001 thrust a more technologically capable and bureaucratically assertive SOCOM into a leading role in a series of regional wars. The early campaigns in Afghanistan and Iraq showcased the new capabilities and compatibility of SOF in the conventional arena. SOF successes in this phase relied heavily on their ability to cooperate with conventional forces to

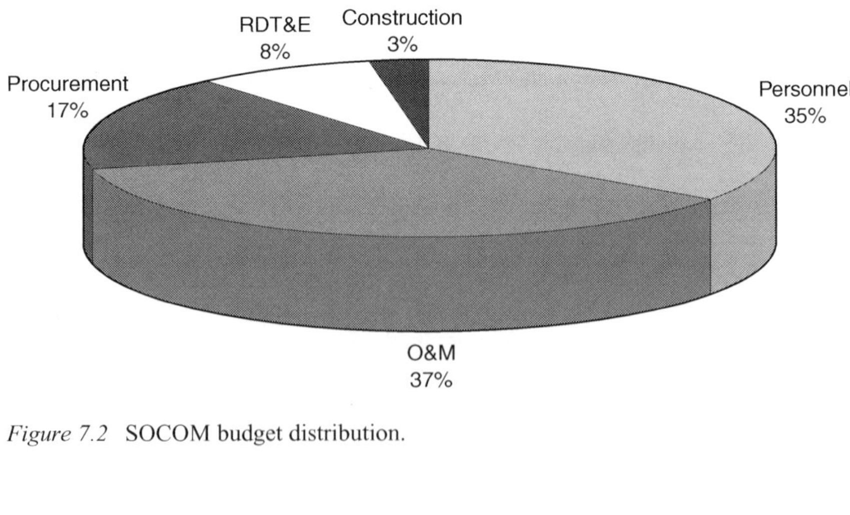

Figure 7.2 SOCOM budget distribution.

defeat conventional enemies at low cost. The follow-on campaigns to pacify Afghanistan and Iraq called for some mix of direct action and unconventional warfare. In both cases, however, the seductive appeal of direct action tended to leave theater commanders short of the unconventional warfare assets needed to train the local forces. Instead of provoking a balanced response from the SOCOM community, the stimulus of irregular warfare amplified existing institutional preferences for direct action.

Afghanistan – major combat operations and the "Afghan" model

After September 11, SOCOM sprang into action along with the CIA's paramilitary wing, SAD. Within weeks of the attack, SAD and SOF (especially Army SF) were working together to aid various groups in Afghanistan (collectively known as the Northern Alliance) against the Al-Qaeda-allied Taliban government. Soon thereafter, SOF began to direct US precision airpower on Taliban forces, resulting in a rapid collapse of the Taliban as the Northern Alliance advanced.[25]

This successful operation, known as Operation Enduring Freedom, was widely hailed as proof positive of the power of "transformation" and the RMA. It was argued that SOF on horseback with laser designators linked to airpower proved decisive with little ground footprint.[26] To some, this image validated Secretary of Defense Donald Rumsfeld's vision of transformation. It also cemented the Secretary's affection for SOF. The "Afghan model" became a new buzzword in defense circles.

Unfortunately, this interpretation of events was flawed. The rapid defeat of the Taliban stemmed from the fact that conventional battle lines already existed in Afghanistan along which the Taliban and Northern Alliance confronted one another in much the same way that combatants on the Western Front in 1915 did. Rather than validating transformation per se, the Afghan model demonstrated that fixed conventional forces that lacked air defense could be subjected to brutal bombardment from modern air forces. This was not a revelation, as it was shown to be true in battles as diverse as the 1968 engagement at Khe Sanh and the 1991 Gulf War. Combined with the pre-existence of Afghan forces that the US could ally with (and in many cases already had a relationship with dating to the anti-Soviet insurgency of the 1980s), the deck was always stacked against the Taliban.[27]

Even as white SOF worked closely with native Afghans, another, covert effort was underway. Black SOF, working with CIA's SAD, established a task force to hunt down high-value targets. Initially known as Task Force Sword, and later Task Force 11 (TF 11), this unit included Delta, SEAL Team 6, and ISA operators as well as British SOF and elements of SAD.[28]

For a variety of reasons, Task Force 11 had mixed results. First and foremost, it was given a very difficult mission of locating elusive enemies on terrain of the enemies' choosing. Other reasons had more to do with its command and composition. Its commander was an Air Force SOF general with little experience with ground force operations. This led to command tensions with other SOF operators in Afghanistan.

Tensions between some members of Delta and Seal Team 6 also emerged during Task Force 11 operations, as did some tensions between white and black SOF. While personal relationships between different operators often overcame these difficulties, it still made operations more challenging. Other tensions emerged between ground component SOF and aviation SOF (both Army and Air Force pilots).

These tensions all culminated in Operation Anaconda in early 2002. An attempt to coordinate US conventional forces and Northern Alliance forces with white and black SOF to destroy what was believed to be a small force of foreign fighters in the Shah-i-Kot Valley, Anaconda became a showcase for poor coordination and intra-SOF rivalry. Though US forces eventually secured their target, the operation exposed the jumbled chains of command and competing jealousies of a host of SOF units eager to claim credit for the last battle of the conventional phase of operations.[29]

Major combat operations in Iraq

As the battles in the Shah-i-Kot raged, plans were being made for the invasion of Iraq. SOF would play several roles in this conflict as well. First, SMUs were primarily tasked with sensitive site exploitation – a special reconnaissance mission involving the investigation of sites thought to contain weapons of mass destruction (WMD). Because no WMD and little evidence existed, this mission turned out to be fruitless.

SOF operators also drove into far western Iraq as part of an economy-of-force mission to inflict damage and pin down Iraqis and disrupt the use of ballistic missiles such as the Scuds that were launched during the first Gulf War. This operation was successful in the narrow sense but had limited overall impact, as there was minimal Iraqi presence in the far west.[30]

Army special operations forces and CIA SAD operators also worked with the Kurds in the north. These SOF forces, collectively known as Task Force Viking, seized an airfield and, in an improvisational move, had C-17 transports bring in armored vehicles to support SOF operations.[31] These armored vehicles provided additional firepower and mobility to the SOF teams, though again resistance was fairly limited.[32] The Navy's SEAL teams were heavily involved in a range of coalition operations in the south, including the actions on the Al Faw peninsula, Basrah, and the advance on Baghdad.

The invasion of Iraq, known as Operation Iraqi Freedom, succeeded far more quickly than its planners had anticipated. This resulted in a vacuum of power into which both conventional and SOF units were thrust. Considerable confusion existed about what SOF should be doing. Some effort was made to help the intelligence community produce a report on Iraqi WMD (known as the Duelfer report) but black SOF quickly found a mission with which it was more comfortable: the hunt for High Value Targets (HVTs).

JSOC formed Task Force 20 (hereafter, TF) to hunt down the key members of the Ba'ath regime including Saddam Hussein and his sons Uday and Qusay. Though the numerical designation has changed many times (to 626, 145, and others), TF has operated continuously in Iraq since 2003.[33] TF's mode of operation has been termed the "unblinking eye," as it uses overhead and other reconnaissance assets to establish persistent surveillance of likely targets. It then conducts "takedowns" of targets at night along with prompt interrogation to allow other targets to be taken that night whenever possible. Further, TF 20 appears to have tremendous autonomy from the normal CENTCOM command arrangements.[34] It appears to report to JSOC, not to the CENTCOM Special Operations Component Commander (known as SOCCENT). This gives it the ability to conduct operations it believes necessary in isolation from conventional units, who actually "own" an area of operations.

According to public sources, the task force has produced mixed results. TF 20 operators were involved in the killing of Uday and Qusay and the capture of Saddam, but it is unclear to what extent their operations provided the intelligence that led to those operations. Some suggest that CIA, white SOF, and even conventional forces may have actually been responsible for the critical intelligence.[35]

After the capture of Saddam and the elimination of the bulk of the "deck of cards" from the former Ba'athist regime, TF 20 focus shifted to other insurgent leaders. Once Al-Qaeda in Iraq emerged as a major component of the insurgency, TF began to focus heavily on it and its leader, Abu Musab al-Zarqawi. Again TF 20 appears to have been part of the operation that led to Zarqawi's death, but it is unclear to what extent their methods provided the intelligence.[36]

SOF acted to support TF 20, and white SOF forces specialized in direct action appear to have joined the task force. At the same time, white SOF practiced unconventional warfare, particularly in Anbar province. There they developed relationships with local tribes who could provide anti-insurgent security.[37] Iraq quickly began to absorb the bulk of the Army white special forces community.

While SOF fought battles in Afghanistan and Iraq, Secretary of Defense Donald Rumsfeld added to its responsibilities. In early 2004, he gave SOCOM the lead in planning and conducting the Global War on Terror, sometimes called the Long War. Rumsfeld also planned to expand SOF as quickly as possible, in part by directing the Marine Corps to participate in SOCOM. MFP-11 money expanded rapidly, nearly doubling, to $7.4 billion by FY 2006.[38]

Counterinsurgency in Afghanistan and Iraq

Even as SOF operated in Iraq, efforts in Afghanistan were ongoing by both black and white SOF as the Taliban re-emerged as an insurgent organization. With extensive sanctuary in Pakistan's lawless western provinces, the Taliban was able to gain some control of southern Afghanistan.[39] At the same time, black SOF was unable to get at either Al-Qaeda or Taliban HVTs, at least on a large scale.[40]

By creating an enormous demand for scarce SOF assets and personnel, the Iraq war limited the involvement of SOF in Afghanistan. US SOF operations in

Afghanistan have been smaller and somewhat lower in profile than operations in Iraq.

The emergence of renewed resistance against US forces in 2003 was a test of SOCOM capabilities and its relationships within and outside the SOF community. While the counterinsurgency mission played to SOCOM and particularly white SOF's traditional strengths, the campaign in Iraq has illustrated the excesses of black SOF autonomy and resources. Though certain white SOF units made concerted efforts to perform a balanced mix of DA, UW, and FID, black SOF's activities caused increasing friction with a series of important stakeholders: white SOF, conventional units (Army and Marine), Iraqi units, and the CIA. The Afghan model had been a showcase of the new compatibility of conventional forces and SOF, but the Iraq COIN model was considerably less encouraging.

The high-value target (HVT) theory of victory

The focus of SOCOM's counterterrorist/counterinsurgency effort has been the so-called high-value target effort.[41] Under the HVT framework, special operators use direct action to capture or kill the leaders of the opposing force. By eliminating the leadership of the terrorist or insurgent group, the operators hope to neutralize or destroy the target organization. The critical activities in the HVT approach are the identification of key leaders, their apprehension/destruction, and the identification of follow-on HVTs. On one level, the HVT theory offers a simple algorithm for the prosecution of operations against irregular opponents; all the command must do is hit the targets on the list and add new names to the list. In the words of then Lieutenant General William G. Boykin, HVT operations in Iraq and Afghanistan are "doing what the Phoenix program was designed to do, without all of the secrecy."[42]

This reference to Phoenix, the mainly CIA effort against the insurgent "infrastructure" in Vietnam is only partly accurate. Instead, the HVT effort has more in common with various decapitation theories of airpower, particularly those developed in the wake of the successful air campaign in Operation Desert Storm. In both airpower and SOF HVT theories, there are critical nodes or individuals whose removal will have outsized effects on the target organization as a whole.

The two theories are not, however, identical. SOF at least makes an effort to capture and exploit HVTs for intelligence rather than simply destroying them. This brings up one of the central differences between the HVT effort and the Phoenix program. The DA arm of Phoenix was composed of Provincial Reconnaissance Units, CIA-sponsored but manned primarily by Vietnamese locals familiar with their area of operations. The Provincial Reconnaissance Units' awareness of the local political and social dynamics made it far easier for them to target enemy cadres and exploit captured insurgents. By contrast, the limited local force involvement in the HVT effort in Iraq and Afghanistan and the rapid turnover in SOCOM personnel (three- to seven-month tours being the current standard) have rendered the HVT approach far less effective than in the latter phases of the Vietnam War.

Real-estate owners vs. shooters

The burden of the occupation duty in Iraq inevitably fell to conventional Army and Marine units. Those units were given responsibility for a series of destructive and constructive tasks. As conventional units made progress in controlling the insurgent threat in their sectors, they shifted from attacking insurgents to maintaining order and building local institutions. Though progress from the "clear" to the "hold/build" phases of counterinsurgency varied considerably by region, it was in the hold/build phases of counterinsurgency that the friction between autonomous black SOF units and American and Iraqi conventional units was greatest. For black SOF operators, the priority was on capturing or killing high-value targets and the primary tool in this effort was the raid. In most instances, raids met two objectives: the elimination of targets and the generation of new intelligence. Suspects detained in black SOF raids were interrogated with the objective of adding new names to the HVT list.

Though few conventional units objected to the basic rationale behind these raids, they bore the cost of collateral damage. Since local commanders often received limited notice of impending raids, they were frequently unable to provide supporting intelligence or warn against problems of false denunciation. Instead, local commands that had worked for weeks or months to build trust with local leaders were often forced to explain the actions of SOF raids to disgruntled local inhabitants and leaders. Tactics that might have been minimally disruptive in the "clear" phase of COIN operations often proved counterproductive once stabilization efforts reached the hold phase.[43]

Conventional units and black SOF also competed for scarce intelligence, surveillance, and reconnaissance (ISR) assets available to support the COIN campaign. Black SOF, invoking the counterterrorism mission, generally absorbed the overwhelming majority of UAV and other ISR assets. Conventional units were forced to make do with the remainder of the available hours to fulfill their high priority missions.

Special operations forces, transformation, and innovation theory

The special operations community underwent greater change in the second interwar period than any of the traditional services. Changes in technology, operational environment, and the bureaucratic landscape drove the capability and status of this previously marginal group to new heights. If the rise of independent air forces was the most striking feature of the first interwar period (1919–1939), the rise of special operations was easily the most striking of the second.

These dramatic changes can be traced to both technology and the international environment. The development of precision weaponry and advanced communications enabled special operations forces to influence major outcomes on the conventional battlefield. These technologies increased the interoperability of special forces and conventional forces. Once treated as marginal and disruptive interlopers, special forces now offered conventional air and ground forces the ability to prosecute the deep battle more effectively. It was this complementarity that encouraged cooperation between the special operations community and conventional communities, particularly the Air Force, that stood to gain capability from these joint ventures.

Environmental changes also increased the importance and status of special operations forces. The European Central Front scenario of the Cold War was gone and with it the focus on high intensity, unlimited war that had long marginalized special operations. The new mission set of limited wars and contingency operations offered special operations far greater opportunity to apply their capabilities. The war on terror in turn provided a clear rationale for expanded direct action and unconventional warfare across a number of regions and individual campaigns. A community that had left the Cold War as a bit player had become a primary actor in the open-ended, irregular campaigns of the twenty-first century.

Though changes in technology and missions increased the capability and standing of special operations, the rise of Special Operations Command owes almost as much to changes in the bureaucratic environment. The failure of the Iranian hostage rescue led civilians to create a special organization, JSOC, to manage future counterterrorism and hostage rescue operations; these efforts culminated in the creation of a four-star, major command with significant funding autonomy and operational authority. These bureaucratic platforms provided the foundation for the dramatic interwar and early GWOT expansion of SOCOM. The September 11 attacks provoked a second bureaucratic shock. The nature and urgency of the threat led the Secretary of Defense to designate SOCOM a “supported” command with umbrella authority for the military campaign against Al-Qaeda worldwide. As the only military service capable of directly attacking the primary American targets in the war, SOCOM was elevated to a leadership role in a war of indefinite duration and scope. If Vietnam had been the first golden age of special forces, and the rise of JSOC in the 1980s the second, then the war on terror had propelled them to a third and even grander period era of prominence.

The rise of SOCOM provides evidence for the three leading theories of military innovation. The bureaucratic changes of the 1980s (JSOC, MFP-11, and SOCOM) and Rumsfeld’s decisions in the early stages of the war on terror are clear examples of civilian intervention. The first set of changes flowed from an episodic failure and a civilian desire to avoid similar embarrassment in the future. In the wake of these Desert One reforms, external threats to American security remained low and civilian attention waned. When the perceived threat to American security spiked after September 11, civilians again intervened and sought to expand the capability and influence of SOCOM through expanded authorities and resources. Eager to produce rapid progress in the war on terror, the Administration went so far as to endorse SOCOM’s preferred strategy of direct action: counterterrorism. This endorsement enhanced the relative position of JSOC and undermined that of the unconventional war community.

Although the rise of SOCOM offers some evidence to support Barry Posen's civilian intervention theory of innovation, there are important qualifications to this interpretation. First, the 1980s intervention appears to have had less to do with expectations of future threats than with bureaucratic embarrassment. The failure of Eagle Claw did not change military and civilian opinions about the relative importance of hostage rescue so much as it revealed a painful inability to perform a highly visible, niche mission. Second, while Secretary Rumsfeld's intervention diverted additional resources to SOCOM, these resource flows did more to increase output than prompt innovation. Additional resources did not trigger a re-examination of theories of direct action; instead they offered JSOC an opportunity to execute existing ideas on a far larger scale.[44] Third, the expanded flow of resources had the unintended consequence of provoking renewed mission and resource competition within SOCOM. While civilians assumed that increased resources for SOCOM would produce a mix of direct action and unconventional warfare capabilities, SOCOM leadership tended to steer most of these funds toward direct action. Members of the white SOF community split into two camps, one competing with JSOC for direct action missions and monies and the other proposing the creation of a separate and co-equal unconventional warfare command.[45]

As these qualifications suggest, there is important evidence to support Stephen Rosen's alternative theory of military led innovation. With the creation of JSOC in the 1980s, the special operations community split into two major factions. JSOC and the black SOF community focused on the problem of direct action while the white SOF community pursued the more traditional missions of unconventional warfare and foreign internal defense. Over the ensuing decades, these two groups developed divergent interpretations of the appropriate employment of special operations forces. While these debates remained largely internal to the special operations community, the terms of this debate had been set by the civilian, bureaucratic interventions of the 1980s. Absent civilian intervention, it is hard to imagine the same intraservice balance of power and schism over roles, missions, and resources.

Competition also plays a powerful role in explaining changes in the special operations community during this period.[46] As noted earlier, one of the unintended byproducts of the Rumsfeld intervention was to provoke competition within SOCOM. Rapid increases in the funds flowing to JSOC increased the relative position of black SOF vis-à-vis the white community. The prospect of diminished relevance and funding led the white SOF community to choose between competition for direct action funds and missions and civilian re-regulation of the division of roles and resources. Competition for direct action missions entailed a narrowing of the functional capabilities of SOCOM as a whole; the "graying" of white SOF could be measured in the diminished emphasis on unconventional warfare. This was particularly ironic as the campaigns in Iraq and Afghanistan highlighted the importance of these same capabilities in counterinsurgency – the training of indigenous forces, military-political engagement with local actors, psychological operations, and civic action. Under these circumstances it is not surprising that some within the unconventional warfare community called for civilian intervention to create a separate organizational structure as a means of restoring balance and limiting the hegemony of JSOC within SOCOM.

The story of the rise of SOCOM is one of dramatic transformation and unintended consequences. A series of changes in technology, contingencies, and bureaucratic structure catapulted SOCOM from a subordinate position into the status of a fifth service. The war on terror cemented the standing and amplified the capability of SOCOM, making it in many ways the focus of an irregular war of indefinite scope and duration. While these changes clearly benefited the special operations community as a bureaucratic entity, their contribution to national capabilities was more mixed. The same set of changes that thrust SOCOM into the lead in the direct action campaigns of the War on Terror led the special operations community away from lower prestige but important, complementary missions of unconventional warfare. It remains to be seen whether the gaps between what civilian policymakers want, need, and receive from SOCOM narrow or widen as the status and autonomy of the command expand.

Notes

1 Hy S. Rothstein, *Afghanistan and the Troubled Future of Unconventional Warfare* (Annapolis: The Naval Institute Press, 2006), pp. 22–23, 92–94.
2 Thomas K. Adams, *US Special Operations Forces in Action: The Challenge of Unconventional Warfare* (New York: Frank Cass, 1998), pp. 210.
3 Susan L. Marquis, *Unconventional Warfare: Rebuilding US Special Operations Forces* (Washington, DC: The Brookings Institution, 1997), pp. 203–204.
4 Michael G. Vickers, "Warfare in 2020: A Primer," Center for Strategic and. Budgetary Assessments, October 1996.
5 Ibid., Rothstein, pp. 134–137.
6 Marquis, *Unconventional Warfare*, pp. 21.
7 Michael Smith, *Killer Elite: The Inside Story of America's Most Secret Special Operations Team* (London: Weidenfeld & Nicholson, 2006), pp. 250–251.
8 Rothstein, *Afghanistan and the Troubled Future of Unconventional Warfare*, pp. xvi.
9 Ibid., pp. 6–7, 10–11.
10 On airpower theories of decapitation, see Robert Pape, *Bombing to Win: Air Power and Coercion in War* (Ithaca: Cornell University Press, 1996), pp. 79–86.
11 Ibid., Marquis, *Unconventional Warfare*, pp. 307.
12 Smith, *Killer Elite*, pp. 264–265.
13 For a detailed account of the pre-Vietnam development of Army special forces see Alfred G. Paddock, Jr., *U.S. Army Special Warfare: Its Origins* (Lawrence: University of Kansas Press, 2002).
14 Adams, *US Special Operations Forces in Action*, p. 74.
15 Jeff McKaughan, "MARSOC Warrior: Delivering Marine Corps Capabilities into the USSOCOM Arena, Interview with Major General Dennis Hejlik," *Special Operations Technology Online*, Vol. 4, No. 5. Online, available at: www.special-operations-technology.com/article.cfm?DocID=1607.
16 For a detailed open source account of ISA, see Michael Smith (op. cit.), *Killer Elite*.
17 There are important limits to this generalization. While the majority of Army special forces are traditionally devoted to UW and FID missions, a number of white units (Army Rangers, the 160th SOAR, and the Navy SEAL teams) are routinely dedicated

to direct action missions. In some cases, these white units supplement black SOF direct action; in other cases they may be employed independently in that role.

18 Fred J. Pushies, *Night Stalkers: 160th Special Ops Aviation Regiment* (Osceola: Zenith Press, 2005), p. 15.
19 Smith, *Killer Elite*, pp. 132–133; Todd Greentree, *Crossroads of Intervention: Insurgency and Counterinsurgency Lessons from Central America* (Westport: Praeger Security international, 2008), pp. 116–121.
20 Marquis, *Unconventional Warfare*, pp. 198–200.
21 Marquis, *Unconventional Warfare*, p. 204.
22 *United States Special Operations Command Posture Statement 2006*, p. 14.
23 Adams, *US Special Operations Forces in Action*, p. 26.
24 Max Weber in H. H. Gerth and C. Wright Mills, *From Max Weber: Essays in Sociology* (New York: Oxford University Press, 1958), p. 233.
25 Benjamin S. Lambeth, *Air Power Against Terror: America's Conduct of Operation Enduring Freedom* (Santa Monica: The RAND Corporation, 2005), MG 166–1, pp. 339–342.
26 Charles Briscoe, Richard Kiper, James Schoder, and Kalev Sepp, *Weapon of Choice: US Army Special Forces in Afghanistan* (Ft. Leavenworth: Combat Studies Institute Press, 2003), pp. 122–125.
27 Stephen Biddle, *Afghanistan and the Future of Warfare: Implications for Army and Defense Policy* (Carlisle Barracks: Strategic Studies Institute, November 2002).
28 Adams, *US Special Operations Forces in Action*, p. 227.
29 The best unclassified account of the operation is Sean Naylor's *Not a Good Day To Die: The Untold Story of Operation Anaconda* (New York: Berkley Books, 2005).
30 Michael R. Gordon and General Bernard E. Trainor, *Cobra II: The Inside Story of the Invasion and Occupation of Iraq* (New York: Pantheon: 2006), pp. 327–334.
31 Adams, *US Special Operations Forces in Action*, pp. 253–254.
32 Major Brian Maddox, "Checkmate on the Northern Front: The Deployment of Task Force 1–63 Armor in Support of Operation Iraqi Freedom," *Armor Magazine* (September–October 2003).
33 Smith, *Killer Elite*, pp. 239–242.
34 Adams, *US Special Operations Forces in Action*, pp. 238–261.
35 Ibid., pp. 238, 259.
36 Frank Scott Douglas, "Waging the Inchoate War: Defining, Fighting and Second-guessing the 'Long War,'" *Journal of Strategic Studies*, Vol. 30, No. 3 (June 2007), pp. 391–420.
37 For a more detailed discussion of the Anbar Awakening and special operations involvement therein, see Austin Long, "The Anbar Awakening," *Survival*, Vol. 50, No. 2 (April–May 2008), pp. 67–94.
38 This figure is the FY 2006 budget request. As such, it does not include supplemental funds for operations in Afghanistan and Iraq. In FY 2005, such supplemental funds added $1.6 billion to the MFP-11 base budget of $6.4 billion. *United States Special Operations Command Posture Statement 2006*, p. 18.
39 Seth G. Jones, "The Rise of Afghanistan's Insurgency: State Failure and Jihad," *International Security*, Vol. 32, No. 4 (Spring 2008), pp. 7–40; Antonio Giustozzi, *Koran, Kalashnikov, and Laptop: the Neo-Taliban Insurgency in Afghanistan* (New York: Columbia University Press, 2007).
40 For a discussion of abortive SOF efforts to pursue targets across the Pakistani border, see Eric Schmitt and Mark Mazzetti, "Bush Said to Give Orders Allowing Raids in Pakistan," *New York Times*, September 11, 2008, and Sean Naylor, "Spec Ops Raids into Pakistan Halted," *Navy Times*, September 28, 2008.
41 *United States Special Operations Command Posture Statement 2006*, p. 5.
42 Quoted in Douglas Jehl and Thom Shanker, "Congress Is Reviewing Pentagon on Intelligence Activities," *New York Times*, February 4, 2005, p. 4.

43 Interviews with Army officers stationed in south and central Iraq, 2004–2006.
44 Ibid., Rothstein, p. 122.
45 Sean D. Naylor, "Support Grows for Standing Up an Unconventional Warfare Command," *Armed Forces Journal* (September 2007).
46 Owen Cote, Jr., *The Politics of Innovative Military Doctrine: The United States Navy and Fleet Ballistic Missiles* (Dissertation, Massachusetts Institute of Technology, 1993); Harvey Sapolsky, "On the Theory of Military Innovation," *Breakthroughs* (Spring 2000).

8 Defense strategy in the 1990s

Old wine, new bottles

Robert R. Tomes

The revolution in military affairs (RMA) identified as a key feature of the second interwar period was in fact the result of decades of Cold War military investments. From the perspective of American defense and military strategy, it is more correct to date the start of the second interwar period from the end of US military involvement in Vietnam. This chapter examines the evolution of American defense strategy in the last decades of the Cold War, arguing that many developments in the 1990s adapted and extended technological and doctrinal innovations began decades earlier without updating the planning assumptions and strategic objectives for their development.

Although the rhetoric changed and the relevant technologies certainly matured, many of the key concepts underlying the American RMA and "transformation" had their roots in a much earlier strategic effort, known as the "offset strategy," to win a conventional war against the Soviets in Europe. Defense strategy in the 1990s was in some sense old wine in new bottles. Ironically, just as the vision of the 1970s began to come to technological and organizational fruition, it proved increasingly ill-suited to the post-Cold War security environment.

At least two possibilities existed for potential innovation in defense strategy. First, strategy could have continued along the path of conventional innovation set by the offset strategy, but in an even more radical way: stressing major efficiencies in force structure to be gained by information technologies, and using such technologies to replicate tasks usually performed by armor and firepower. In short, it could have produced a significantly smaller, lighter, faster, and more lethal force in conventional operations that relied on unmanned systems, that had intelligence and civil affairs capabilities that could support any regional conflict, and that extended operational capabilities into urban environments.

Second, defense strategy could have responded to the environmental stimuli of the second interwar period: it could have attempted to integrate the information and targeting advantages of the offset strategy into a small wars environment. It could have used the lessons of American operations in the 1990s to revamp and revitalize service doctrine for a task much different from a conventional fight in Cold War Europe.

Neither innovation occurred. Instead, OSD retreated behind a set of RMA jargon that – if it reflected anything – reflected the same ideas animating the offset strategy: increased situational awareness and information management for the purposes of precision targeting. This failure to innovate in strategy has many fathers, but can be seen especially clearly in two factors: there was little civilian effort, attention, or intervention into strategy debates at a high level; and the services had little interest in pursuing innovation.

This chapter details the ideas behind the offset strategy in the 1970s and 1980s, and then traces their fate in the 1990s and beyond. It does so in four parts. First, the offset strategy is described, and similarities are noted with contemporary rhetoric. Second, the offset strategy's three major components are examined, with the conclusion that a real innovation in strategy occurred during the Cold War. The third section contrasts this innovation with the impoverishment of strategic discourse during the interwar period. A final section details the emergence of the language of "transformation" during the late stages of the second interwar period and the disruption of transformation planning in the post-9/11 period. The chapter concludes by reviewing what when wrong and what might be possible in the near term.

The post-Vietnam offset strategy

Historian Alex Roland notes that the "Army concluded from Vietnam that it needed not less technology but more. It was not that smart weapons were bad; rather, they were not smart enough."[1] General William C. Westmoreland was among those who had already begun to think in earnest about how emerging technology would shape future battles. In early 1969, as American support for Vietnam plummeted, he outlined a vision in which "enemy forces will be located, tracked, and targeted almost instantaneously through the use of data links, computer assisted intelligence evaluation, and automated fire control."[2] Using language nearly indistinguishable from later defense transformation, Westmoreland envisioned "twenty-four-hour real- or near-real-time surveillance of all types" and forces possessing area control systems exploiting "communications, sensors, fire direction, and the required automatic data processing."[3]

A year later Congressional hearings on the "electronic battlefield" concluded that "the electronic or automated battlefield represented a whole series of technologies and programs that were combining to form a totally new American way of war."[4] Senator Barry Goldwater opined that battlefield information systems represented "the greatest step forward in warfare since gunpowder."[5] The stage was set for a revolution in American military forces. Richard Swain notes that, as the Army sought to revise its doctrine, training, and force structure after Vietnam, the vision Westmorland articulated in 1969 became the "new mantra." Army General William DePuy echoed this notion in 1974, predicting that in the future, "What can be seen, can be hit. What can be hit, can be killed."[6] For current students of defense planning and military thought, there is nothing new in this Cold War vision – but for the defense planners of the 1970s and 1980s, it was a seemingly unobtainable goal, very different from the emphasis on nuclear strategy that had previously dominated defense planning.

By the late 1970s, then-Secretary of Defense Harold Brown and Undersecretary of Defense for Research and Engineering William Perry had conceived of an "offset strategy" to address strategic and operational challenges posed by Soviet conventional forces. These forces were steadily growing more capable. American defense planners feared that a rapid armored assault would easily penetrate NATO's defenses and cripple its tactical nuclear forces. Once Soviet forces seized NATO territory, Moscow could hold the West hostage to demands. In this scenario, NATO's nuclear escalation doctrine, including the threat of nuclear defense, were no longer viable or credible. NATO would not use nuclear weapons on NATO territory to unseat the Red Army.

The offset strategy was designed to exploit the technological revolution described above. In brief, superior Russian numbers would be "offset" by superior American technology. NATO forces would use surveillance and precision technology to find and kill follow-on echelons of Russian troops before they reached the central front. To be effective the offset strategy would require: improved sensors, for locating Russian forces; improved computing technology, for analyzing different kinds of sensor data; improved communications, for transferring data to shooters; and improved precision weapons, for hitting targets accurately. In short, it would require a revolution in military affairs. But if the offset strategy worked, DePuy's vision would be realized – NATO would find, hit, and kill Russian forces before they overwhelmed the defensive line.

Although not formalized as a DoD program in the same guise as flexible response or massive retaliation, programs pursued under the offset strategy, including the Assault Breaker concept demonstration, stealth aircraft, and communications advances, nonetheless shaped modernization decisions in the late 1970s and became the de facto principle underlying the Reagan defense build-up in the early 1980s. Perry recognized that experimental activities already underway at DARPA could increase the overall effectiveness of US conventional forces. In 1978 testimony to the Senate Committee on Armed Services, he outlined why precision strike advances offered the " 'greatest single potential for force multiplication" to meet the Soviet threat in Europe:

> Precision-guided weapons, I believe, have the potential for revolutionizing warfare. More importantly, if we effectively exploit the lead we have in this field, we can greatly enhance our ability to deter war without having to compete tank for tank, missile for missile, with the Soviet Union … In sum, the objective of our precision-guided weapon systems is to give us the following capabilities: to be able to see all high-value targets on the battlefield at any time; to be able to make a direct hit on any targets we can see; and to be able to destroy any target we can hit.[7]

Perry later observed that the essence of this offset strategy, which "sought to use technology as an equalizer or 'force multiplier,' " was in fact "pursued consistently by five administrations during the 1970s and 1980s."[8]

For retired admiral Bill Owens, too, the capabilities labeled revolutionary in the early 1990s derived from operational approaches and systems "engineered and acquired in the late 1970s through the late 1980s," which made victory in the 1991 Gulf War "inevitable and our historically small loss of life probable."[9] Former Assistant Secretary of Defense for International Security Policy Ashton Carter adds that after the offset strategy's precepts were "dramatically demonstrated during Operation Desert Storm" they became "key to Washington's way of waging war."[10] Indeed, Richard Van Atta and Michael Lippitz argue that a broadly defined approach for leveraging emerging technology had emerged, in theory at least, by the end of the 1970s.

Substantively, the offset strategy was organized around the idea that several technology projects could be simultaneously matured and integrated into a larger system, a task that further brought systems engineering, systems integration, and information technology into the emerging vision for defeating successive waves of Soviet armored echelons. The offset strategy leveraged developments in sensors, new weapons platforms, and advanced munitions "to greatly mitigate or potentially negate the Soviet threat and do so without resort to the use of nuclear weapons."[11]

The offset strategy was also an early expression of what current defense planners call "system of systems" thinking. Through a "synergistic application of improved technologies" that spanned "electronic countermeasures, command and control (communications, data links, and networks), stealth, embedded computers (microprocessors), and precision guidance (advanced sensors)," the offset strategy would "allow the US to overcome Soviet defenses and destroy Soviet tank legions."[12]

In sum, the language of the offset strategy bears a striking resemblance to that used to describe defense strategy in the second interwar period. This resemblance is no accident. In the next section, the three principle achievements of the offset strategy are described: the Assault Breaker concept, improvements in battlefield intelligence, and training and doctrinal changes. These represent a real innovation in defense strategy. However, rather than building on this innovation, strategic thought in the second interwar period stagnated, simply replicating the ideas of the previous generation.

Assault Breaker: seeing, hitting, and killing

The Assault Breaker concept demonstration was central to the offset strategy and one of the Cold War's most ambitious systems integration efforts. Assault Breaker, a multifaceted effort combining sensors, weapons systems, weapons platforms, and information-processing tools aimed to test Perry's vision: to see all high-value targets on the battlefield at any time; to be able to make a direct hit on any targets that could be seen; and to be able to destroy any targets that could be hit. "The question to be answered," Richard Van Atta relates, "was whether developments in sensors, computing, communications, guidance, and munitions allowed for deep precision attack against hard, mobile targets."[13]

Although some of the program's ambitious goals remained unrealized until the early 2000s, by the 1980s Assault Breaker had demonstrated "a capability to attack multiple tank targets using terminally guided submunitions released from a standoff 'missile bus' controlled by airborne radar."[14] Among the other elements of the larger system for testing the concept was an attack coordination center to fuse data from multiple sensors. Testing of key Assault Breaker components began in fiscal years 1979 and 1980, with critical demonstrations in the early 1980s. In late 1982, for example, the above-mentioned guided submunitions "made direct hits, one on each tank in a pattern of five stationary tanks."[15] Assault Breaker reflected the growing "realization that the timely use of tactical nuclear weapons to stop an attack in [Europe] was unrealistic."[16]

Service cooperation was critical. As Van Atta, Nunn, and Cook conclude, perhaps "even more important than the testing and developing of specific technologies is the conceptual breakthrough in getting the services to work together across the barriers of roles and missions to attack the Warsaw Pact tank threat."[17] Assault Breaker's management structure was based on a joint program office that struggled to address often diverging interests of service sponsors. Army and Air Force collaboration was in part facilitated by the support of Generals Don Starry, commander of the Army's Training and Doctrine Command, and William Creech, head of the Air Force Tactical Air Command. Additional Army and Air Force support was due to the fact that Assault Breaker did not threaten the discontinuation of either service's existing programs.[18]

Interestingly, the first combat testing of the essential conceptual and technological elements of Assault Breaker occurred in the Persian Gulf, not on the plains of Central Europe. During the Gulf War, for example, some thirty-two Army Tactical Missile System missiles were used in conjunction with targeting information provided by the Joint Target Surveillance and Attack Radar System aircraft. These operations mimicked important aspects of the Assault Breaker demonstration. As Paul Nitze remarked about advanced conventional weapons developed for the defense of Europe, the "Gulf War offered a spectacular demonstration of the potential effectiveness of smart weapons used in a strategic role."[19]

Information advances in the 1970s and 1980s

Another part of the offset strategy, battlefield intelligence, also began a transformation process in the 1970s that culminated in the 2000s with persistent theater surveillance capabilities. A 1974 Army Intelligence Organization and Stationing study documented many of the problems plaguing battlefield intelligence in the early 1970s. The study "found that military intelligence units were not properly organized to support the tactical mission, and, indeed, were in most cases beyond the control of tactical commanders because of their strategic responsibilities."[20] In today's parlance, the information given by battlefield intelligence was not "actionable."

Many of the innovations undertaken in the 1970s aimed to correct this problem, laying the foundation for American military dominance and helping shape visions of future warfighting capabilities that dominated Cold War military thought, such as information superiority, dominant battlespace awareness, decision superiority, persistent surveillance, and strategic pre-emption. Two programs contributing to advanced battlefield information capabilities were the Coherent Emitter Location Testbed (CELT) and the Battlefield Exploitation and Target Acquisition (BETA) initiative. Both contributed to the evolution of American battlefield information superiority by demonstrating, at least in experiments, that intelligence and information systems could be made more relevant to the commander.

CELT, "the first automatic, near-real-time system for precision location of communications emitters," was demonstrated during NATO exercises from 1978 through the 1980s.[21] Antecedents included 1960s initiatives to locate North Vietnamese electronic emissions, the Air Force–DARPA Emitter Location System project in the mid-1970s, and the Precision Location Strike System from the same era. Responding to increased concerns with Soviet conventional forces, the Emitter Location System was renamed CELT in 1978 and focused on developing "automatic location and classification" of vast numbers of enemy "emitters expected on the European battlefield, with the accuracy required for targeting by standoff weapons."[22] After successfully demonstrating the ability to locate emitters and generate information for targeting, CELT technology and operational concepts would contribute to subsequent systems, informing how sensor and targeting systems could be linked to fire control and strike systems.

Among the lessons applied to later systems included "the need for a disciplined systems engineering approach to future developments," a conclusion that reinforced the emphasis on integrating air and ground forces.

Related systems included the Joint Tactical Information System and Joint Tactical Fusion Program. The Joint Tactical Information System provided a secure capability to move data around on the battlefield. Less vulnerable to enemy jamming, it worked by spreading data transmissions over different frequencies. The Joint Tactical Fusion Program merged a secure communications capability with visualization tools able to represent the fusion of data. Together they aimed to increase situation awareness by providing secure, jam-resistant, near-real-time information updates to commanders.

This was the first time that commanders were provided an integrated, near-real-time picture of their battlefields that was secure enough to share information from classified sources. Since then, the "common operational picture" has become almost an end in and of itself, with more and more data pushed to tactical forces that increasingly lack the expertise and resources to digest, understand, and act on what they are receiving.

Training, doctrine, and vision in the 1970s and 1980s

Technology was only a part of the story, though. Military leaders embraced organizational innovation in the late 1970s. Preparing to actually fight – and perhaps win – a conventional conflict impelled new thinking about deep strike for a generation of leaders raised on the shallow logic of nuclear strategy and doctrine. Planners reconsidered maneuver and interdiction. Improvements in training, revisions of doctrine, and new operational concepts provided the essential integration mechanisms through which new technology was brought into the larger military innovation process.

An overhaul of Army training significantly improved battlefield performance. Indeed, training – not technology – appears now to be the critical variable in the revolutionary leap in effectiveness ascribed to US military forces. The National Training Center (NTC) at Fort Irwin, California, was formed to provide the level of realistic and challenging training needed to prepare for the fight in Europe and remains the world's most sophisticated armored warfare battle laboratory for experimenting with new concepts and weapons.

In addition to training unit commanders to think more creatively, NTC provided a testing ground for the "big five" core weapons systems of the revitalized Army: the M1 Abrams main battle tank, the M2 Bradley armored infantry fighting vehicle, the AH-64 Apache combat helicopter, the UH-60 Blackhawk utility helicopter, and the Patriot air defense missile system. Linking them together was a new arsenal of information and command and control systems that multiplied the baseline capabilities of individual weapons systems.

Although many problems remained, including cross-service interoperability challenges, important advances in intelligence, surveillance, and reconnaissance transpired. A revolution occurred in America's overall command and control capabilities. Weapons systems integrated and enabled by information technology led to operations being more lethal over greater distances with far fewer forces. Institute for Defense Analyses researchers Richard Van Atta and Michael Lippitz concluded that "the ability to exercise military control [*shifted*] from forces with the best or the most individual weapon systems toward forces with better information and greater ability to plan, coordinate, and accurately attack."

Other developments in the 1980s set the stage for advances in intelligence support to military operations. These included more timely and precise information for target identification and location. Better information for precision strikes opened additional routes to develop smart weapons with computerized guidance and navigation systems. The automated information fusion and decision tools developed as part of the offset strategy and Assault Breaker program spurred additional efforts to expedite command and control of battlefield strikes and to integrate additional systems into a larger, more capable command and control network. Military planners adapted their thinking about modernization and systems development. The traditional focus on single weapons platforms or individual systems shifted to a new focus on how different platforms and systems worked in concert. Integration, interoperability, and then "jointness" successfully ascended in the defense planning and strategy lexicons.

The early joint information integration projects discussed above evolved into the common operating picture and geospatial awareness capabilities central to defense transformation in the 2000s. *In toto*, these changes were remarkable. The American military evolved into a better integrated, better trained, more technologically savvy force; one that very likely could have succeeded in holding off a Soviet invasion of Western Europe. The advances in surveillance, intelligence, and targeting were very real, and organizational and doctrinal change occurred to implement them. The next section compares the success of the offset strategy with what followed.

The second interwar period

Since the 1991–1992 Gulf War, US strategic planning and military thought have been inundated with arguments that American armed forces exhibit a significant, discontinuous increase in military effectiveness, "the process by which armed forces convert resources into fighting power."[23] Andrew Marshall, Director of the Department of Defense's Office of Net Assessment, assigned the first official study of the changes in military affairs attributed to American forces after the first Gulf War to Lieutenant Colonel Andrew F. Krepinevich, then a military aide in his office. Although new technologies like those showcased in the Gulf War did suggest a new era in military affairs, Krepinevich concluded that the more important development for the future of warfare involved a new approach to "system integration and organizational innovation."[24]

What type of organizational innovation did Krepinevich have in mind when he wrote of changes in military affairs? It was not the kind of innovation that would lead to the organizational reforms typically associated with fundamental changes in military affairs or defense strategy. Rather, the ability to integrate new information quickly and achieve unity of purpose across land, sea, and air forces facilitated a new level of sophistication in command and control during combat operations, but seemingly obviated the need for fundamental changes in the peacetime organizational structure. Thus, the potential for the creative destruction of innovation was undermined, and the stage set for more of the same. Throughout the 1990s, defense analysts touted the potential for information and intelligence systems to provide unprecedented, shared situational awareness – the exact accomplishment of the offset strategy.

The problem was that the very nature and philosophy of the system integration and organizational innovation that underwrote the American RMA focused attention on the so-called emergent properties or multiplier effects produced by linking together existing systems. Of course, new doctrine and operational command structures emphasizing joint force integration and operations were later grafted onto existing service or component structures in order to actually leverage new capabilities. But the focus of discussions was integration of existing technology, creating false expectations about the current abilities of the established organizations. Instead of pushing forward to new ideas and new organizational structures, the strategic discourse focused on bringing together what the military already had.

Muted in the defense strategy discussion of the 1990s was the essential underpinning of any information superiority regime – intelligence beyond the battlefield. Defense planners seized on the simplistic notions of information superiority that emerged from earlier discussions of capabilities developed during the 1970s and 1980s. That is, they focused on situational awareness, battlefield monitoring, and the false promise of a common operational picture. As even Krepinevich noted, the idea of information dominance may be "relevant to all levels of conflict, from the grand strategic to the tactics," but in the ideal situation, "it is established in peacetime and sustained in pre-crisis and crisis periods, and in war."[25] Information dominance requires deep knowledge of the adversary, the operating environment, and the nature of the potential conflict, not just data points representing targets on a computer screen. This was all the more true in an era of limited rather than total wars.

In the 1990s, defense strategists and pundits merely grafted info-speak terms and concepts onto the forces and capabilities developed under the offset strategy. It is instructive to recall the operational origins of RMA terms like information dominance, information superiority, and decision superiority. "Information dominance," for example, was defined in the early 1990s "as a superior understanding of a (potential) adversary's military, political, social, and economic structures, to include their strengths, weaknesses, locations, and degrees of interdependence, while denying an adversary similar information on friendly assets."[26] By the 2000s, however, this deeper meaning of information dominance and other terms seemed all but forgotten, with the ideal vision reduced to a purely technological common operational picture. The importance of other forms of intelligence was neglected. The potential for innovation was replaced by the situational awareness/targeting complex begat by the offset strategy.

The effect of the international environment was also ignored. Operationally, the post-Cold War Army gained what should have been invaluable experience in regional conflicts that focused attention on communications, situational awareness, and indirect fire needs. Somalia offered a lesson in the stark reality of urban combat against an enemy that could not be distinguished from the civilian population. It was an enemy aided by radical Islamists teaching local militiamen to down American helicopters with rocket-propelled grenades, a tactic battle-tested against Russian forces in Afghanistan a decade earlier. Meanwhile, the Air Force employed precision strike capabilities during the 1990s in diverse combat environments. Doing so exposed inherent limitations of systems and targeting processes conceived for the battlefields of Cold War Europe. Operation Allied Force in Kosovo surfaced apparent shortcomings in heavy lift, logistics, and intelligence support to military operations. Failure to find Serbian armor in Kosovo, or to prosecute timely attacks after they were located, revealed gaps in the "kill chain" that questioned the effectiveness of precision air strikes in small wars and urban settings.

But none of this experience affected the strategic debate. From the perspective of strategic innovation, how weapons are employed and the organizational and operational innovations that enable their effectiveness are the important dimensions of "change" one must consider. Through the offset strategy, the military, and especially the Army, had developed new doctrine, training regimes, and technology that supported the new operational imperative to seize and maintain the initiative in the highly dynamic and increasingly intense battlefield environment suggested by the Yom Kippur war. It was a strategy aimed at developing the ability to fight outnumbered and win the opening battles of a war in Europe without resorting to nuclear weapons use. But by the mid-1980s, the same offensive approach had come to guide Army planning for non-European contingencies, regardless of whether it was appropriate to those scenarios or not. Every thing became a lesser-included case of the "big war," whether it was urban fighting, counter-insurgency operations, or peacekeeping.

Although the underlying capabilities labeled an RMA retained their revolutionary nature when viewed through the lens of military history, and although defense intellectuals benefited from RMA theory as a framework for thinking about historic changes in military affairs, much of the RMA talk devolved into empty rhetoric. Many forces were driving this lack of strategic change. An "old wine, new bottles" syndrome inhered in many programs because bureaucratic entrenchment and infighting among the services. Douglas MacGregor, an Army officer who led thinking on the need for a reorganization of post-Cold War ground forces, joined the critics. Reflecting on 1990s defense modernization efforts, he lamented that defense modernization discourse resulted in mere "bumper stickers" that did "not prevent competing service requirements from dominating joint integration efforts."[27] Indeed, it seemed the services liked the offset strategy just fine.

The inattention of the executive branch also contributed to intellectual stagnation. Successive planning activities and documents record the lack of interest in furthering the spirit of military innovation that had dominated modernization planning in the late 1970s and early 1980s. The 1993 *Report of the Bottom-Up Review*, the 1994 *Nuclear Posture Review*, the 1994 Commission on the Roles and Missions of the Armed Forces, and the 1997 *Report of the Quadrennial Defense Review* all failed to be "decisive in setting clear guidance" or establishing "a consensus for policy objectives."[28] "In each of these cases," American foreign policy scholar Janne Nolan concludes, "senior leaders, beginning with the president, proved reluctant to engage the issues directly or to provide leadership to guide the outcome."[29]

This lack of civilian inattention could be seen in military plans. Major defense planning documents, including the three cited above, drew heavily from Cold War planning assumptions. A significant cross-border armored attack remained the chief planning scenario driving force-allocation requirements despite the marked decline in the mechanized forces of potential enemies. Planners framed regional instability and political violence as lesser, wholly included scenarios that forces designed for cross-border contingencies could cope with. Rapid dominance and rapid deployment concepts from Persian Gulf contingency planning were reinforced.

Domestic politics provided severe disincentives to push OSD strategy toward innovation. Political and fiscal realities were in fact ill-suited to furthering the innovation paths begun in earlier decades. Modernization efforts were delayed, with funding directed to redressing critical readiness issues associated with successive military interventions abroad. Department of Defense planning guidance for 1994 and 1995, moreover, anticipated savings of some twenty-five billion dollars from cutting force structure and reducing support infrastructure. The General Accounting Office "blasted these assumptions" in 1996, arguing "there is no significant infrastructure savings."[30] Modernization funding continued to suffer, leading analysts to conclude in 1998 that "defense procurement" was "down by more than 70 percent since its high point in the mid-1980s" and "billions below the requirements to recapitalize America's defense forces."[31]

An absence of organizational urgency for transformation in part reflected the lack of a compelling political reason to embrace it. Voters seemed not concerned about the level of defense spending. Despite a renaissance in American military thought, and perhaps because new thinking was not affecting change evenly across mission areas or quickly enough within services, Secretary of Defense William Cohen noted "a sense of intellectual and doctrinal stagnation" among some military leaders.[32] For these and other reasons, the 1990s are best considered a transition period in which leaders did not or could not seize opportunities for additional defense transformation. This does not mean RMA discussions were altogether feckless. As Colin Gray argues, "The raising of the RMA flag mobilized a wide variety of perspectives and skills" and "enabled some long-antecedent ideas and streams of analysis to play significantly in a contemporary debate."[33] Indeed, they set the stage for later policy discussions and provided much of the language used to frame a new transformation strategy.

Transformation: everything old is new again

The term "transformation" was present within but not central to 1990s defense policy discourse. References to transformation during much of the decade appeared primarily within RMA discussions about historically profound changes in military history or as a descriptor for specific leaps in the effectiveness of one or more combat arms. The term's ascent in defense planning discourse toward the end of the decade was facilitated by a number of widely cited sources, some warranting mention here to gain insight into how defense planning discourse evolved. The key question remains: was this shift in discourse also a shift in strategy?

The early evidence suggested it was not. Among the benchmarks in the shift toward transformation dialogue was the 1997 National Defense Panel report *Transforming Defense – National Security in the 21st Century*. Around this time, defense analysts and military theorists began addressing the issue of defense transformation and calling for a renewed planning debate. The National Defense Panel report represented an evolution in thinking among defense interlocutors about the pace and scope of defense modernization. Its report argued for initiatives to leverage information technology, to develop additional space-based capabilities, and to accelerate organizational innovations. Initially, "transformation" seemed to represent incremental implementation of the offset strategy.

However, the late 1990s witnessed a potential shift. The 1998 Department of Defense *Annual Report to the President and the Congress*, through which the Secretary of Defense communicates the status of defense readiness and planning, was the second document marking the shift from RMA language to transformation processes.[34] Previous annual reports contained rather pedestrian discussions of RMAs as historical phenomena, using the RMA construct to conceptualize ongoing changes in the nature of war. Of the five main sections of the 1998 report, one addressed service transformation and another department-wide transformation. Additional text was devoted to "New Operational Concepts" and "Implementation" needs, including experimentations, demonstrations, and other activities required to facilitate a larger transformation effort.[35] This could have been the precursor to an approach of the scope and scale of the offset strategy, but moving beyond it.

This trend continued as subsequent annual reports addressed transformation strategy in lieu of the previous editions' RMA chapters. The switch represented the acceptance of "transformation" as a term of art within defense planning and policy circles. To many, it suggested that something more prescriptive than existing RMA language was needed. For example, more forceful transformation language was included in a September 1999 Defense Science Board (DSB) report entitled "DoD Warfighting Transformation." The DSB defined transformation as "a process that seeks fundamental change in how an enterprise conducts its business" in pursuit of "discontinuous change in the nation's capabilities to conduct" military operations.[36] Departing from the deterministic and exogenous view of organizational change inherent in some RMA discussions, transformation was defined as a "self-inflicted" process seeking "very big change."[37]

Among the few, muted references to RMAs in the DSB report was the recognition that "very big change" was "sometimes characterized by the term 'revolution in military affairs'."[38] Apparently, rather than accepting the passivity of an "RMA-is-certain" approach, the DSB articulated a view of defense reform and modernization that sought an alternative to RMA-associated rhetoric. Instead of questing after immediate, revolutionary reforms, transformation was characterized as "defining and implementing a vision of the future different from the one embedded, if only implicitly, in DoD's current plans and programs."[39] Here, at last, was the beginning of a real "conventional" innovation: a new vision, requiring large organizational change, that took advantage of RMA technologies but tried to move beyond the offset strategy.

But, like in the early 1990s, the same obstacles remained. Service disinterest and executive apathy once again thwarted strategic change. The DSB was among a number of quasi-official defense policy organizations that "did not find much sense of urgency" for significant change in service warfighting capabilities. The DSB, for example, concluded that "the focus and effort needed" to transform was being underestimated.[40] Despite concerted efforts within the Office of the Secretary of Defense to push modernization, the Clinton administration did not make transformation a priority in terms of leadership attention, a willingness to expend political capital to influence Service decisions, or a clear vision for change conjoined with "sticks" to induce compliance. This left the military services to define their own visions without an overarching mandate to change – or to integrate. Defense transformation did not seem to be getting any traction.

The language of transformation may have actually made matters worse. The climate of reform and fascination with RMAs and "re-engineering" in Washington gave the appearance that change was widespread. Champions of the defense status quo actually used the 1991 Gulf War to rationalize the continuation of long-standing organizational structures and weapons systems, rather than as an impetus to true innovation. There was no shift in the prioritization of technology research and development efforts. This was still the case in September 1999 when then presidential candidate George W. Bush promoted his vision of defense transformation.

The pace and scope of US defense modernization was scrutinized during and after the 2000 election. Eliot Cohen recounted the essence of the ensuing criticism. Strategically, American defense strategy was too wedded to "a Cold War-derived understanding of military power" and failed to "focus on the challenges of the new century: homeland defense, a rising China, and what can only be termed 'imperial policing'."[41] Even after a decade of RMA defense discourse and numerous visions for future warfighting, technology development and procurement processes adhered to "Cold War paths," leaving "systems suited for a war in Europe with the defunct Soviet Union rather than hardware optimized for" emerging threats.[42] Former defense official Ashton Carter urged the incoming Bush administration to transform defense and to revamp research and development. The United States, he argued, was "not fully exploiting or staying abreast of the information revolution."[43] It was merely continuing with the offset strategy.

The evidence from the early Bush administration was mixed. Soon after assuming office in 2001, the George W. Bush administration convened several panels to chart a new course for US defense modernization. One chaired by Andy Marshall, the Director of the Office of Net Assessment, was tasked to rethink and revitalize efforts to modernize the US military. Ostensibly, Marshall returned to first instincts, in this case the approach taken in the early 1990s when the idea of an RMA took root among defense planners. Once again, the Office of Net Assessment sponsored studies to explore changes in warfare. These studies benefited, however, from a decade of thinking about changes in how warfare should be waged in the information age. Among their themes were arguments for lighter, more lethal forces able to befuddle opponents with rapid dominance. Many believed that armored forces would not be needed once units were fully enabled by information technology and ever more advanced knowledge-to-action capabilities. Troops would be wrapped in a protective layer of information dominance.

The Transformation Study Group, as the panel was titled, provided its findings to Secretary of Defense Donald Rumsfeld in late April 2001. The report supported the institutionalization of a new defense transformation process, describing the process as potentially facilitating "changes in the concepts, organization, process, technology application, and equipment through which significant gains in operational effectiveness, operating efficiencies, and/or cost reductions are achieved."[44] But the administration was not ready to implement the report's findings. It postponed changing defense programs until it could complete a strategic planning process that fully considered the risks involved in such changes. Much uncertainty remained in terms of how fast defense transformation could proceed. By the end of the first George W. Bush administration, "transformation" was adopted as an umbrella term for attempts to remake US ground forces in the model of a lighter, more agile, information-enabled precision force wielding greater lethality over greater distances in less time. But OSD had shown little interest in actually implementing the strategy in action.

No prominent changes occurred during the Bush administration's first year in office. Critics resorted to citing Bush's own campaign speeches lamenting that the American military was "still organized more for Cold War threats than for the challenges of the new century – for industrial-age operations, rather than information age battles."[45] They asked when he would live up his promise to correct what he called the "the last seven years" of "inertia and idle talk."[46]

Defense transformation plans were themselves transformed by the 9/11 terrorist attacks on the World Trade Center and the Pentagon and, subsequently, by the global war on terrorism. Indeed, in a November 1 *Washington Post* op-ed, Secretary of Defense Donald Rumsfeld referred to the attacks on American soil as "a wake-up call" that created a "new sense of urgency" for modernizing and transforming the Armed Forces. The bottom line: "Transformation cannot wait."[47] Similarly, then-Undersecretary of Defense for Acquisition and Technology Pete Aldridge viewed the war on terrorism as creating "a springboard to transformation" and as stimulating the impetus to overcome the "status quo."[48]

Transformation assumed the tone of a strategic imperative in the December 2001 Department of Defense *Annual Report to the President and the Congress*. The language of the report implied an accelerated pace and broadened scope, although the administration announced it would delay making significant programmatic changes until all the newly commissioned transformation studies were completed, fully analyzed, and utilized to inform a new defense transformation strategy. Underlying the report was a clear message: the global war on terrorism would not delay transformation.

Yet the opposite has occurred. The global war on terrorism and the continued occupation of Iraq have ensured that little real transformation will transpire in the coming decade. Moreover, post-9/11 operational experiences have revealed the superficiality of many of the changes that did occur in the 1990s. Many of the technologies envisioned as revolutionary in the 1970s and 1980s have come to fruition, but they have not been followed by needed organizational changes. This reality, combined with a security environment that is inherently less amenable to technological solutions designed for standoff battles, marks a strange irony in the arc of defense transformation since the end of Vietnam, with the second interwar period merely the midpoint of this arc, rather than the beginning or the end.

Conclusion

This chapter has shown how defense strategy turned from innovation to relative stagnation in terms of adapting core capabilities and achieving the same overwhelming dominance in other areas of warfare. The offset strategy of the 1970s and 1980s was a highly innovative set of organizational changes that had real effects on the ground. By contrast, the "RMA" and "transformation" debates in the defense community during the second interwar period represented little more than a rehashing of the offset strategy, moving to vision from a theater-based contingency plan to a global way of warfare that assumed enemies would not themselves adapt.

Defense strategy was neither conventionally innovative, by taking technological development to the next level, nor environmentally innovative, by responding to the missions of the 1990s. Instead, it remained focused on the same problem – a Cold War style conventional war – and proposed the same solution: increased situational awareness and targeting ability. But this increase in awareness and targeting had its roots in the threat posed by the Soviet Army. Rethinking the "enemy" and the style of warfare would not come until after the US was bogged down in Iraq and was forced to relearn counter-insurgency warfare.

The causes of this stagnation are manifold, but two deserve to be singled out in connection with the theories of innovation noted in the introduction. First, there was little civilian intervention in defense strategy in terms of innovation and rethinking the types of warfare the US would face. High level policy-makers were generally inattentive, and when attentive, were not willing to either push big ideas or back them with political will. Civilians by and large ignored the burgeoning potential for innovation during the Clinton administration, and approached the issue with extreme caution under Bush. Even the shock of 9/11 could not move defense strategy: transformation was declared essential, but there was little move to either a conventional or an environmental change.

Second, there was almost no interest among the military services for change in defense policy. Each service mostly preferred to continue with its current funding priorities, and decided not to rock the boat. There are few examples of innovative strategies being pushed under the protection of senior officers. There services did not compete with one another to offer strategies that emphasized their own priorities while attacking those of other services; instead they colluded to preserve the status quo. In general, the services were quite happy with what the offset strategy had wrought, and saw no reason to risk their gains for improvement.

In the end, then, defense strategy innovation was not achieved in the second interwar period. The question remains: with environmental and budget pressures stronger than ever, will any of the latent causes of innovation be brought to the fore? For sure the military will continue to adapt Cold War forces to fight and win small wars and to prosecute "stability" operations, missions that were not a focus for planners during the second interwar period. Many of the information systems and "awareness" capabilities developed and adapted in the last decade provide a suitable starting point for innovation. But the expensive, decades-long acquisition systems that are legacies of the offset strategy remain at the core of Service programs. When the budgets contract to balance the budget, it remains to be seen whether the Services will cut systems largely conceived for bygone wars or pursue innovations on the magnitude of what occurred in the 1970s and 1980s.

Notes

1 Alex Roland, "Technology and War," *American Diplomacy* (Vol. II, No. 2), posted at www.unc.edu/depts/diplomat/AD_issues/4amdipl.html.
2 Westmoreland cited in M. W. Hoag, *New Weaponry and Defending Europe: Some General Considerations* (Santa Monica, CA: The Rand Corporation, October 1973), p. 13.
3 Westmoreland articulated this vision in a 1969 speech to the Association of the US Army, reprinted in Appendix A of Paul Dickson's *The Electronic Battlefield* (Bloomington, IN: Indiana University Press, 1976), pp. 215–223.
4 Ibid., pp. 116–117.
5 Ibid., p. 101.
6 General William E. DePuy correspondence to Gen. Creighton W. Abrams, January 14, 1974, in Col. Richard M. Swain (ed.), *Selected Papers of General William E. DePuy, USA Retired* (Fort Leavenworth, KS: Combat Studies Institute, 1994), p. 71. Additional background on TRADOC and the DePuy cite is also cited in Swain.
7 Perry cited in Richard Van Atta, Jack Nunn, and Alethia Cook, "Assault Breaker," in Richard Van Atta and Michael J. Lippitz, with Jasper C. Lupo, Robert B. Mahoney, and Jack Nunn, *Transformation and Transition: DARPA's Role in Fostering an Emerging Revolution in Military Affairs* (Alexandria, VA: Institute for Defense Analysis, 2003).
8 William J. Perry, "Desert Storm and Deterrence," *Foreign Affairs* (Vol. 70, No. 4), p. 69.
9 William Owens with Edward Offley, *Lifting the Fog of War* (New York, NY: Farrar, Straus and Giroux, 2000), p. 89.
10 Ashton B. Carter, "Keeping America's Military Edge," *Foreign Affairs* (January/February 2001), p. 99.
11 Ibid.
12 Richard H. Van Atta and Michael J. Lipptz, *DARPA's Role in Fostering and Emerging Revolution in Military Affairs* (Alexandria, VA: Institute for Defense Analyses, November 19, 2001), p. 7.
13 Ibid., pp. iv–14.
14 Van Atta *et al.*, *DARPA Technical Accomplishments Vol. II*, pp. 5–17.
15 Ibid., p. iv–18.
16 Van Atta *et al.*, Assault Breaker chapter (2003), p. iv–1.
17 Van Atta *et al.*, "Assault Breaker," pp. iv–9, iv–41.
18 Van Atta *et al.*, *DARPA Technical Accomplishments Vol. II*, pp. 5–1.
19 Paul H. Nitze, "Is It Time To Junk Our Nukes?" *Washington Post* (January 16, 1994), p. C1.
20 Kenneth Allard, *Command, Control, and the Common Defense* (New Haven, CT: Yale University Press, 1990), p. 147.
21 Van Atta *et al.*, *DARPA Technical Accomplishments Vol. II*, pp. 7–1.
22 Ibid.
23 Allan Millet, Williamson Murray, and Kenneth Watman, "The Effectiveness of Military Organizations," in Allan R. Millet and Williamson Murray, *Military Effectiveness, Volume I: The First World War* (Boston, MA: Unwin Hyman, 1988), p. 2.

24 Andrew F. Krepinevich, Jr., *The Military-Technical Revolution: A Preliminary Assessment* (Washington, DC: Center for Strategic and Budgetary Assessments, 2002), p. 8.
25 Ibid., p. 22.
26 Ibid., p. 22.
27 Douglas A. MacGregor, "The Joint Force: A Decade, No Progress," *Joint Force Quarterly* (Winter 2000–2001), p. 20.
28 Janne E. Nolan, *An Elusive Consensus: Nuclear Weapons and American Security After the Cold War* (Washington, DC: Brookings Institution Press, 1999), p. 86.
29 Nolan, *An Elusive Consensus*, p. 86.
30 Robert W. Gaskin, "A Revolution for the Millennium," in Williamson Murray, ed., *The Emerging Strategic Environment: Challenges of the Twenty-First Century* (Westport, CT: Praeger, 1999), p. 133.
31 Ibid.
32 Ibid., p. 56.
33 Gray, *Strategy For Chaos: Revolutions in Military Affairs and the Evidence on History* (London: Frank Cass, 2002), p. 17.
34 *Annual Report to the President and Congress* (Washington, DC: Department of Defense, 1999), Chapters 13–15.
35 Ibid.
36 Ted Gold, *Report of the Defense Science Board Task Force on DoD Warfighting Transformation* (Washington, DC: Office of the Undersecretary of Defense, 1999), p. 3.
37 Ibid.
38 Ibid.
39 Ibid., p. 30.
40 Ibid., p. 1.
41 Eliot A. Cohen, "Defending America in the Twenty-First Century," *Foreign Affairs* (Vol. 79, No. 6), p. 41 [40–56].
42 Ibid., p. 41.
43 Ashton B. Carter, "Keeping America's Military Edge," *Foreign Affairs* (January/February 2001), p. 92.
44 Department of Defense Transformation Study Group, "Transforming Military Operational Capabilities," online, available at: www.defenselink.mil.news.Nov20001/t11272001_t1127ceb.html, p. 5.
45 Andrew J. Bacevich citing Bush in *American Empire: The Realities and Consequences of U.S. Diplomacy* (Cambridge, MA: Harvard University Press, 2002), p. 140.
46 Ibid.
47 Donald H. Rumsfeld, "Beyond This War on Terrorism," *Washington Post*, November 1, 2001, p. 35.
48 Quoted in Adam J. Hebert, "Aldridge: War on Terrorism Demands Major Changes in Acquisition Practices," *InsideDefense.com*, October 31, 2001.

9 The RMA and the defense industry

Eugene Gholz

The discussion of American military innovation in the 1990s held out the prospect of substantial changes in the way that each of the military services would fight future wars. Capitalizing on advances in information technology, the military ideally would be able to decentralize operations (massing fires instead of forces), choose precise targets (minimizing the scale of destruction and especially collateral damage), and optimize strikes for political effect (integrating intelligence databases with tactical operations). If fully implemented, the tasks facing leaders and soldiers during both wartime and peacetime would change.

Apostles of the revolution in military affairs (RMA) often downplayed the role of systems acquisition in the transformation effort. They argued that the main changes would be in organizational culture and training, affecting the way the military fights rather than what it fights with.[1] They hoped that the transformation project would not get bogged down in decades-long technology development efforts – efforts whose expense might make the revolution seem unobtainable. Moreover, innovation in commercial information technology (IT) visibly progressed at a blistering pace, and the price of civilian computing capacity plummeted every year. Technology seemed to be the easy part of the RMA.

Some new equipment would of course be needed to add communications and information processing capability to existing platforms – to enable soldiers, sailors, and airmen to share the information that would allow them to fight differently. But at first glance, the core military platforms could stay the same, and the IT acquisition would be relatively easy. The services need only reach out and grab what they wanted from the incredible commercial market.

Even if pure "off-the-shelf" buying rarely seemed reasonable, the whole point of IT seemed to be its adaptability, and few commercial businesses were immune to the whirlwind. So why should military buyers not also be able to get customized hardware and software for their needs with little effort? The acquisition side of the transformation effort would draw from a new set of suppliers with expertise in IT, whether established commercial suppliers or new start-up firms.[2] These firms rather than the traditional defense industry would do the bulk of the relatively small development and manufacturing task for the RMA.

But this relatively limited effort would not take full advantage of the RMA: different platforms, built from the ground up to depend on networks for their full functionality, would be smaller, faster, simpler, and cheaper. If a platform could reach out through the network to draw on other assets for part of its capability – for example, if a sensor did not itself need to take a shot but could complete the "kill chain" by communicating with a separate "shooter" platform – then the platform would need to pack fewer systems into its confined space, and its form, fit, and function could be optimized for a single purpose.[3]

This new development trajectory implied a threat to the established defense industry, which is composed of firms expert in making high-end, multi-function equipment. The core competencies of the defense industrial base, according to this line of thinking, were quite different from those of the information age. Proponents of military transformation sometimes talked about replacing the existing set of suppliers: in the early 1990s, they introduced the idea of the RMA with an analogy to Wal-Mart that emphasized information technology's ability to drive down costs and increase competitiveness, and then they progressed to describing Wal-Mart's effect on established retailers, driving K-Mart and Sears to the brink.[4] New companies replaced the old in the commercial market; the business side of the RMA, supplying equipment to the military, was perhaps even more naturally susceptible to creative destruction than the military services themselves were. After all, Schumpeter developed the idea to describe technological progress in industry.

The intuition was that the established defense industry might present a barrier to the RMA, because traditional firms would try to foist the same old products on the military or because those firms lacked the core competencies in IT.[5] If this were true, the watchful, far-sighted buyer – the committed transformation agents in the Department of Defense – would have to consciously drive acquisition to start-ups and commercial IT firms, driving out traditional defense firms. The policy conversation generally did not criticize the "Iron Triangle" or the "Military-Industrial Complex" so openly; the American national security leadership (civilian and uniformed) is used to defending itself against political opponents who use those phrases. Nevertheless, the Office of the Deputy Undersecretary of Defense for Industrial Policy set about creating lists of "transformational" technologies, seeking out new suppliers with whom to build new relationships.[6] The result would be the expected "destruction" in the industrial landscape.

The reality of acquisition in the 1990s and early 2000s was quite different. The military needs the established defense industry, and the buyer had neither the interest nor the capability to drive the big players out of business. Lockheed Martin, Boeing, General Dynamics, Northrop Grumman, and Raytheon led the traditional defense industry in adapting to the RMA, finding ways to work with the buyer to create "transformational" systems after all – or at least to create marketing materials that capitalized on the language of the RMA.[7] In fact, almost every planned defense acquisition program developed a set of briefing slides to explain its essential relationship to transformation, and those slides were actually grounded in the defense industry's relatively good understanding (compared to commercial suppliers and start-ups) of the military's innovative intentions and new, IT-driven requirements.[8]

Not all of the industry's adaptations were good for defense acquisition – specifically, the industry willingly cooperated with the radical expansion of the "systems approach" to acquisition, including its most extreme form, "Lead Systems Integrator" mega-contracts that over-promised technological progress and multiplied political inefficiencies in defense management. But the system served the transformation effort. The military got funding for a new generation of equipment, as ever pushing the technological envelope even if the contractors could not deliver on their most extreme promises of high performance, prompt delivery, and affordable cost.

The customer–supplier relationship

Well-functioning organizations are built around a few core ideas.[9] If an organization is too big or tries to do too much, its managers and employees cannot prioritize their efforts and develop day-to-day standard operating procedures that achieve a level of functional performance sufficient to sustain the organization's access to resources. Specifically, government organizations (including the military services and their acquisition components) need to do enough of what Congress and interest groups want that they avoid scandals, investigations, and oversight that take away their decision-making authority and budget appropriations. Firms in the defense industry that live off government contracts need to reassure their military customers that they can deliver a technological edge to American fighting forces even as they convince Congressional appropriators that they are doing so in a way consistent with American capitalist values.

The challenge of bureaucratic politics, including in the defense sector, is for organizations' leaders – their executives – to build relationships with key constituencies. Military leaders constantly need to reassure Congress, politicians, and pundits that they understand their role in American grand strategy and that they are building and training the most effective fighting force possible to serve the American national interest. Defense industry executives specialize in understanding military needs and the strategic language in which they are expressed – terms quite unlike those used by the exclusively profit-oriented customers of commercial firms.

For the defense industry, the customers' operational concerns depend on military concepts that are alien to most technological entrepreneurs and businessmen.[10] While the normal relationship between an innovative supplier and a customer must combine shared understanding of technical needs and capabilities with shared understanding of value-added and economic constraints, the defense market imposes the additional demand for the supplier to share the customers' understanding of military operations. The military has a bewildering array of organizations involved in planning, technology development, and acquisition, and the defense community is famous for its jargon and acronym-ridden communications. Only specialists can figure out what the various organizations' products mean.

But the distinct relationship between the military and the defense industry extends beyond their shared ability to interpret the jargon in requests for proposals.

The military needs a trusted partner with whom to discuss half-baked ideas about evolving doctrine. The companies have built that trust over the years by making understanding military operations one of their true critical tasks. They track the relevant documents, attend the important conferences, send managers to participate in working groups with their customers, and hire retired military officers, who bring both personal operational experience and a network of informal contacts and friendships with officers still on active duty. Over time, the defense firms' business development staffs have developed the ability to judge which draft documents will have lasting influence, separating them from the bulk of the military reports and proposals that are ephemeral or insignificant. They also advise warfighters on which possible doctrinal changes would stretch technological capabilities farther than even the understandably optimistic business development people think attainable. Each firm's military-facing business development staff serves as the interface between the warfighters and the companies' technical experts.

Commercial firms and start-ups lack that crucial connection to the military customer. Moreover, they are also unprepared for the non-economic evaluation methods through which military contracts are awarded. Even if they could follow the operational requirements in sufficient detail, they would still struggle to adhere to defense contracting requirements that try to achieve social goals. When the Congressional representatives who actually pay for defense acquisition think of "American capitalist values," they demand a particular style of competitive bidding, and they insist on preferences for subcontracting to small and minority-owned businesses. Those preferences are a long-standing characteristic of the American political system, even though they are inconsistent with the laser-like focus on efficiency that most commercial customers demand.[11] Few suppliers can develop standard operating procedures to work in both commercial and defense acquisition environments.

Furthermore, commercial suppliers also struggle to comply with complex accounting rules that give the military customer insight into high-risk technology development efforts and enable the buyer to (relatively) confidently offer cost-plus development contracts. Countless small IT service providers have sprung up in the commercial market – e.g., firms that help companies use open-source Linux as the operating system for their intranets or firms that write custom applications through which particular clients can use Oracle's database platforms – but the military has been unable to hire a similar set of companies for its needs. The military market is too small to induce the large commercial IT firms (Microsoft and Oracle, for example) to customize their own products for the unusual military operational environment. But at the same time, it is prohibitively difficult for boutique start-ups to clear the security hurdles of working on military IT systems and to deal with other unique Defense Department administrative overhead.

The result is that larger, established firms dedicated to the defense IT environment (e.g., SAIC) are booming, and IT services are one of the fastest growing components of revenue and profits for the traditional defense prime contractors (e.g., Northrop Grumman and General Dynamics have each acquired a number of medium-sized military information technology and software support firms such that they are now big players in the field). Throughout history, prime contractors have often bought start-up companies to connect the technological innovators with the primes' understanding of the military customer, perhaps most notably during the "electronics revolution" in the 1960s.[12]

The defense prime contractors also subcontract to buy commercial IT equipment and software, even if the commercial firms do not sell directly to the military. Indeed, prime contractors routinely manage subcontracting networks to get access to particular inputs or technologies, ranging from commodity materials to specially designed high-end components. The National Steel and Shipbuilding Company (NASSCO, now owned by General Dynamics) is one of the nation's "Big Six" core military shipbuilders but, despite its name, it has never produced its own steel. More closely connected to the IT-RMA, prime contractors can write "middleware" to adapt commercial IT for military customers, but they can also develop custom applications in-house. Defense prime contractors are good at deciding between those options: they understand the customer's needs in a way that commercial firms cannot, and they analyze alternative ways to fulfill those needs – their bread-and-butter "make-buy decisions."

During the 1990s, defense prime contractors sought to expand by capitalizing on that "systems integration" capability.[13] If they defined their core competency as the ability to manage complex acquisition projects, then there were few natural limits to the breadth of projects that they could take on. Their promises fell on a bipartisan consensus of receptive ears: the Democratic administration sought to "reinvent government" to emphasize efficiency, and the Republican "revolution" in Congress strove to replace "sclerotic" civil servants with "nimble" private-sector managers whenever possible.[14] The government outsourced management of big defense acquisition projects to prime contractors with Lead Systems Integrator contracts, winnowing the defense acquisition workforce (particularly cutting science and technology experts) and reinforcing the dominance of the established defense industry prime contractors.[15]

The RMA, because it relies so intensely on cross-platform communications and protocols that enable data fusion, naturally offered ammunition to those who argued that single contracts should encompass multiple platforms, allegedly to enable smooth integration of their designs. In the public eye, systems integration often boiled down to giving a single firm responsibility for ensuring interoperability throughout the force structure – specifically by making that firm responsible for delivery of an enormous variety of products.[16] The RMA matched centralization on the buyers' side with centralization among the private sector firms of the defense industry: prime contractors' influence if anything grew, either through acquisition to give them sufficient breadth of technical capabilities or through their roles as management hubs for networks or teams of suppliers.

Perhaps military leaders should have objected to giving up so much control over key platforms and the architecture of the overall "system of systems." In the long run, the design and features of major equipment influence organizational culture, and the services have a strong interest in protecting their organizational essence.[17] But in the 1990s, warfighters generally trusted their long-standing relationships with their friends in the defense industry more than they trusted their relationship to what they perceived as less-responsive civil-servant acquisition specialists. Indeed, while the Joint Chiefs and operational commanders almost always liked high-tech weapons – the American military is hardly techno-phobic – warfighters' relationships with the technically informed parts of the services are rarely comfortable. Military officers with engineering specialties are limited by a glass ceiling compared to their counterparts in the combat arms.[18] Meanwhile, at lower levels of the hierarchy, the military asks little from its laboratories beyond operational testing and last-minute kludges to prepare systems for deployments; laboratory executives garner little respect from "real" warfighters.[19] As a result, the RMA reinforced the comfortable organizational relationship between the military and the established defense prime contractors. Information technology-intensive transformation may have created new capabilities without destroying much of the old in the defense industry.

The politics of acquisition investment

Even if the military wanted to buy new systems directly from nontraditional suppliers – that is, discounting the organizational politics that work against creative destruction – Congressional politics would have posed a nearly insurmountable barrier. Unlike a commercial enterprise, the military does not generate its own investment funds (or raise them from capitalists); government agencies get their money through Congressional appropriations. The military needs the lobbying power of the industry to convince Congress to extract resources from taxpayers (or perhaps more realistically to convince Congress to borrow money, committing future politicians to extract resources from taxpayers).

Congress has many other ideas about how to spend money.[20] In the American system of checks and balances, Congress weighs the military's natural desire to provide as much protection as possible against requests to fund education, welfare, and other programs. Especially during the "peace dividend" decade of the 1990s, representatives might have downplayed military proposals for new systems acquisition – after all, the Soviet threat had collapsed, and while it might have seemed reasonable to request some new types of equipment for the post-Cold War security environment, it was hard to agree on a major investment initiative without a visible, widely agreed threat to national security.[21] Acquisition decisions in the 1990s required backing from civilian lobbyists combined with Congress' respect for professional military.

The traditional defense industry cultivates lobbying expertise. Just as the established defense firms' business development offices monitor and understand the firms' military customers, their Washington lobbying offices' full-time job is to follow the twists and turns of Congressional support for defense programs. The promise of technology like the IT-RMA can capture the imagination of political leaders, but on the other hand, unless an innovation promises to respond to a specific strategic problem, legislators will also ask how it will affect the interests of key constituents in their Congressional districts. The established defense industry's lobbyists live to answer that question.

Investment in the IT-RMA – especially for the original vision that emphasized doctrinal change based only on limited spending on network equipment rather than on the development of new platforms – was constrained by the relatively weak political support for spending on communications equipment. The defense acquisition budget has traditionally been organized on a platform-by-platform basis, making it hard to buy a lot of new RMA-linked information technology. The services like this platform-centric acquisition system, because it gives their acquisition commands ready control over the projects that can influence the core identity of service branches. Congress likes this acquisition system, too, because buying platforms fits legislators' political preference to allocate large sums of money to high-profile programs for which politicians can claim credit.[22] The facilities that assemble fighter aircraft and warships each employ thousands of workers in a particular location, so the ups and downs of platform acquisition in the defense budget have huge effects on the local business cycle – and on local politicians' prospects for re-election. By contrast, information-technology projects tend to be produced in smaller, less capital-intensive facilities with much less political visibility.

However, the IT-RMA changed the political constellation. Network-centric warfare calls for bigger investment than ever before in communications equipment, which will attract more industrial players, and it apparently requires a new level of technical cooperation among firms to develop standards and protocols, embodied in new industry organizations like the Network-Centric Operations Industry Consortium. It is only a small step from a consortium that allows firms to discuss conceptual approaches and technical standards to a consortium that helps its members overcome competitive pressures and cooperate on lobbying strategy.[23]

Lobbying power also contributed to the adoption of Lead Systems Integrator contracting. Politically weak organizations led the charge, because the new approach promised to respond to their political problems. For example, the Army, working with its traditional supplier base, lost its struggle to fund the Crusader artillery system. Its next major proposal, the Future Combat Systems, promised to develop new indirect-fire weapons simultaneously with new direct-fire weapons, reconnaissance platforms, utility vehicles, and other equipment. That combined proposal could of course draw on a broad spectrum of defense industry lobbyists, and the need for integrated communications and network-centric applications justified the coalition.

Perhaps more importantly, the acquisition concept copied a "successful" acquisition model pioneered by the Missile Defense Agency, which has consistently received strong funding support despite nearly religious ideological opposition from some political factions. The new model also brought a larger, more influential aerospace firm into the fold as an Army supplier when the Boeing-SAIC team won the integration contract. But the new prime contractors did not squeeze out the old: the Army's traditional suppliers like United Defense and General

Dynamics each got a substantial share of the planned production work.[24] The information technology backbone of the project simply provided the framework through which the Army strengthened political support for its acquisition initiative.

Similarly, the Navy added Lockheed Martin as lead integrator on one of its Littoral Combat Ship designs (and Raytheon as the prime contractor for its new amphibious ship class, the LPD-17), broadening support beyond the traditional shipbuilding sector. And the Coast Guard's Deepwater program consciously sought political capital from its leadership team of defense industry prime contractors (Lockheed Martin and Northrop Grumman) but still found plenty of work for its traditional small Gulf Coast and Great Lakes shipyards.[25] Again, the projects emphasized the political advantages of creation without destruction.

Conclusion

The bottom line is simple. Politics (Congressional and organizational) saved traditional platforms and firms from the threat posed by the RMA in the 1990s. But politics also enabled tremendous spending on new technology. As part of the investment in RMA technologies, the military shifted to the Lead Systems Integrator approach, concentrating huge projects in the hands of a few, centralized teams. In this model, all of the buyer's eggs are in one basket. The expensive investment efforts have not run smoothly, and the customer has found it difficult to find alternative approaches or to adjust when a contractor runs into technical or management problems.[26] Major "transformational" investments in the Army's Future Combat Systems, the Navy's Littoral Combat Ship, Air Force space systems, and the Coast Guard's Deepwater program have been suspended or exempted from rules designed to protect against cost overruns and performance shortfalls for reasons of "national security" – code for the dearth of reasonable alternatives.

There is little doubt that the military will eventually get a new generation of first-line equipment and that the new equipment will rely on information technology to a far greater extent than previous generations. But non-traditional suppliers will not lead the development of that new generation of equipment. New products will be created, but not with Schumpeter's expected industrial destruction.

Notes

1 "DoD IW Experts Outline What They Need from Industry," *Aerospace Daily*, June 8, 1995, p. 384; Malina Brown, "Cebrowski: Transformation Initially Should Skip Acquisition Realm," *Inside the Navy*, November 25, 2002.

2 Vago Muradian, "DoD-Industry Both Must Change to Quickly Meet Transformation Needs," *Defense Daily International*, February 22, 2002.

3 Vice Admiral A. K. Cebrowski and Captain Wayne P. Hughes, Jr., "Rebalancing the Fleet," *US Naval Institute Proceedings*, Vol. 125, No. 11 (November 1999), pp. 31–35; Hunter Keeter, "Business As Usual Won't Allow Navy to Achieve Its Net Centric Vision," *Defense Daily*, April 3, 2001.

4 Vice Admiral Arthur K. Cebrowski and John J. Garstka, "Network-Centric Warfare: Its Origin and Future," *US Naval Institute Proceedings*, Vol. 124, No. 1 (January 1998), pp. 28–35; Frank Wolfe, "Navy Wants to Engage Industry on 'Front End,' Official Says," *Defense Daily*, August 19, 1998.
5 "'Revolution in Military Affairs' Could Pose Problems for US Military," *Aerospace Daily*, October 21, 1996, p. 107.
6 For a description of the expected transformation of the defense industry as a whole, see Office of the Deputy Undersecretary of Defense (Industrial Policy), *Transforming the Defense Industrial Base: A Roadmap*, February, 2003. That report was followed by a number of follow-on studies, each of which emphasized a particular part of the defense industry. See, for example, Office of the Deputy Undersecretary of Defense (Industrial Policy), *Defense Industrial Base Capabilities Study: Command and Control*, June, 2004. The Office of Force Transformation also worked with venture capital firms to try to stimulate entry of new defense suppliers. Bruce V. Bigelow, "New War, New Allies," *San Diego Union-Tribune*, November 26, 2002; Jefferson Morris, "Transformation Office to Partner with Venture Capital Firms," *Aerospace Daily*, September 30, 2002, p. 3.
7 Companies began advertising themselves as "network-centric," and they began to justify mergers and acquisitions because they helped create "system-of-systems" companies, explicitly adapting the language of the RMA to describe their business goals. See, for example, Rich Tuttle, "JTRS Win Fits System-of-Systems Approach, Boeing Executive Says," *Aerospace Daily*, June 27, 2002, p. 6; Vago Muradian, "Kresa: TRW Deal Would Make Northrop Key Space, 'Transformation' Player," *Defense Daily International*, March 1, 2002.
8 Congress made it clear that projects designated as "transformational" might receive some political protection. Hunter Keeter, "Transformation Turns Up Heat on Programs, Officials Say," *Defense Daily*, September 6, 2002.
9 James Q. Wilson, *Bureaucracy: What Government Agencies Do and Why They Do It* (New York: Basic Books, 1989), pp. 91–101.
10 Peter Dombrowski and Eugene Gholz, *Buying Military Transformation: Technological Innovation and the Defense Industry* (New York: Columbia University Press, 2006), pp. 24–25.
11 Thomas L. McNaugher, *New Weapons, Old Politics: America's Military Procurement Muddle* (Washington, DC: Brookings Institution, 1989), p. 124; Harvey M. Sapolsky, "Equipping the Armed Forces," *Armed Forces & Society*, Vol. 14, No. 1 (Fall 1987), pp. 113–128.
12 G. R. Simonson, "Missiles and Creative Destruction in the American Aircraft Industry, 1956–1961," in G. R. Simonson, ed., *The History of the American Aircraft Industry: An Anthology* (Cambridge: MIT Press, 1968), pp. 228–241; Davis Dyer, *TRW: Pioneering Technology and Innovation since 1990* (Boston: Harvard Business School Press, 1998), pp. 218–221.
13 Eugene Gholz, "Systems Integration in the US Defense Industry: Who Does It and Why Is It Important?" in Andrea Prencipe, Andrew Davies, and Michael Hobday, eds., *The Business of Systems Integration* (New York: Oxford University Press, 2003), pp. 281–284, 286–287.
14 For an example of Congressional interest in privatizing defense management – and how proponents of military transformation used that interest – see Vago Muradian, "Defense Panel Considers Unified Procurement Organization," *Defense Daily*, February 9, 1998.
15 Valerie Bailey Grasso, *Defense Acquisition: Use of Lead Systems Integrators (LSIs) – Background, Oversight Issues, and Options for Congress*, Congressional Research Service Report RS22631, March 26, 2007.
16 Gholz, "Systems Integration," p. 280.
17 Carl Builder, *The Masks of War* (Baltimore: The Johns Hopkins University Press, 1989), pp. 22–24, 112–114.

18 Perhaps a seeming counter-example proves the rule: ADM Rickover drew power and longevity from his special relationship with Congress.
19 Dombrowski and Gholz, *Buying Military Transformation*, pp. 116–118; Harvey M. Sapolsky, *Science and the Navy: The History of the Office of Naval Research* (Princeton: Princeton University Press, 1990), pp. 86, 89, 96–98.
20 Harvey M. Sapolsky, Eugene Gholz, and Allen Kaufman, "Security Lessons from the Cold War," *Foreign Affairs*, Vol. 78, No. 4 (July/August 1999), pp. 77–89.
21 Eugene Gholz and Harvey M. Sapolsky, "Restructuring the US Defense Industry," *International Security*, Vol. 24, No. 3 (Winter 1999–2000), pp. 5–51.
22 Dombrowski and Gholz, *Buying Military Transformation*, p. 109; McNaugher, *New Weapons, Old Politics*, pp. 15, 135–137.
23 Dombrowski and Gholz, *Buying Military Transformation*, p. 110.
24 Nick Jonson, "FCS Competitors Likely to Receive Future Work, Boeing Says," *Aerospace Daily*, March 11, 2002, p. 4; Pat Towell, "Appropriators Cut Army's Crusader, But Add Money to Save Local Jobs," *Congressional Quarterly Weekly*, June 28, 2002; Emily Hsu, "GD, United Defense Team Up for Army's Future Combat Systems," *Inside the Army*, January 27, 2003.
25 Vikram Mansharamani, *The Deepwater Program: A Case Study in Organizational Transformation Inspired by the Parallel Interaction of Internal and External Core Groups* (SM thesis, Massachusetts Institute of Technology, 2004).
26 Eugene Gholz, "Czars and Company Men: The Wrong Way to Run National Security Space Acquisition," *Georgetown Journal of International Affairs*, Vol. 5, No. 2 (Summer/Fall, 2004), pp. 11–19.

10 The RMA and the second interwar period

Harvey M. Sapolsky, Brendan Rittenhouse Green, and Benjamin H. Friedman

This book has surveyed the efforts of America's civilian and military national security bureaucracies to adjust to the technological advances of the revolution in military affairs (RMA) and the demands of the new post-Cold War era. Our chapters on the armed services show technological evolution combined with a failure to transform organizations to meet the new threat – that is, there was creation *without* destruction.

How the US military ought to have changed since the Cold War is controversial. That it has not changed much should not be. Today, even amid two counterinsurgency campaigns, the services' preferences have barely budged from the days of Fulda Gap and the Northern Flank. The platform names have changed and their capability has increased, but the story is the same. The Navy wants more ships and has designs in the works for bigger and better frigates, destroyers, cruisers, carriers, and amphibious and support ships. The Army wants to refurbish equipment taxed by use in Afghanistan and Iraq, and bring into the force new systems capable of destroying the motorized rifle regiments and guard tank divisions of the future. The Air Force is worried about the aging of its aircraft and wants to expand the buy of the F-22 and acquire new tankers. The UAV era is upon us, but mostly to identify targets and assess battle damage. Both the Marines and Special Forces have new equipment on order for all their old missions. Conventional war capabilities increased, but conventional transformation never occurred.

The counterinsurgency revolution never came either. All of the services' conventional obsessions turned out to have little applicability to the small wars they ended up fighting in the second interwar period. The marines and the army made some small adjustments to doctrine, and some minor innovations were made in the use of technology, but generally considered counterinsurgency a lesser-included case. When the services were forced to do counterinsurgency, it became clear that environmental transformation had not occurred.

More generally, the major organizational changes one would expect from transformation were simply absent. From the end of the post-Cold War drawdown in 1998 until the end of the second interwar period, there was only a modest decrease in personnel and a virtual stasis in force structure. Even American nuclear forces, which underwent a severe reduction in role in national strategy, exhibited little change. This continuity illustrates the lack of transformation.[1]

Here we explain why the US military has not changed more. To draw together the many influences leading to this result that were alluded to in the previous chapters, we first sketch a political picture of the second interwar period. It is one with a relatively peaceful international environment; ad hoc military and foreign policy changes; a corresponding disjunction between the conflicts in which America is involved and the military it has to fight them; and a military that embraces technology and is uncomfortable with counterinsurgency.

We then apply theories of organizational politics and innovation to this landscape to explain why there has not been more innovation. In the second interwar period, the foundations of innovation were rarely observed: civilians remained quiescent, organizational innovators could not find protectors, and the services log-rolled rather than competing. The result was some non-trivial changes, but nothing transformational. We conclude that the future promises to look like the recent past.

A long peace?

The Cold War ended with the collapse of the Soviet Union. When the Soviet Union broke up, nearly all of its East European allies and component republics defected to the Western alliance, and its Communist ideology was universally discredited and abandoned. The few Cold War issues not settled by the Soviet Union's collapse were unable to hold international interest and soon fell from the status of being considered major problems. Simultaneous with the end of the Cold War, the United States defeated a direct military challenger, in the form of the Iraqi seizure of Kuwait. The force created to defeat the Soviet Union easily defeated the Iraqis and liberated Kuwait in one of history's most lopsided victories. Not only was the Cold War over, but there seemed to be no lingering military problems that the United States had not already solved. The United States and the capitalist system had triumphed; leaving no doubt that America was the dominant world power, militarily, economically, and culturally.[2]

Many thought at the time that the 1990s were to be the beginning of a turning away from international military confrontation, a long peace. Europe began to focus inwardly on its own integration and the absorption of nations to its East. Russia tried to come to grips with its devastated economy and disoriented polity. Japan sought to cope with a lingering recession that destroyed the vision that it was the next global giant. After suppressing a democratic awakening among its youth, China reconciled economic vibrancy with one-party rule, at least temporarily. And the rest of the world, no longer dominated by the US/USSR competition, turned to local concerns.

Everywhere defense budgets were cut. As the surviving superpower, the United States had plenty of room to reduce defense expenditures, but cut less than its allies did. Republicans, and especially President George H. W. Bush, fought to keep the cuts limited by preventing direct transfers of dollars saved in defense to increase social program expenditures as many Democrats desired.[3]

New technology and new wars

There were far more military interventions by American forces in the decade after the Cold War than in the twenty-five years prior to its conclusion.[4] There could be, because interventions no longer risked confrontation with the Soviet Union. Russia might occasionally object to a particular intervention, but few cared about its interests or threats. A superpower no longer lived there, to borrow a phrase of the times.

The technology of warfare also changed. Stealth, precision weapons, satellite communications, and UAVs began to play important and much heralded roles in US military operations. Starting with the Gulf War, the so-called Revolution in Military Affairs (RMA), which had been much discussed in the military literature in the 1980s, became part of the public discussion of US military capabilities. As discussed in Chapter 2, the RMA can trace its origins back to World War II, when the search for technological advantage grew to be a central component in the American military effort. Sixty years of roughly $60 billion of annual investments seeking ways to substitute American science and engineering for American soldiers were paying off.[5] As the Gulf War demonstrated, US forces had the technology and the training to defeat large, if somewhat inept, militaries in conventional warfare quickly and with minimum US casualties.

The lack of true rivals and technological advancement interacted to change the nature of warfare for the United States. Because casualties could be kept low and there would be no superpower challenge to our action, American leaders began to take seriously past policy statements about the need for the US to play a central role maintaining global stability. There was no big threat to counter the American military, so the think-tank and specialist postulating about future missions and force structures included much discussion of peacekeeping, ethnic conflict, democratization, and nation-building tasks. The military did not much like these options, preferring instead to cite Iraq and North Korea as major security concerns with the knowledge that they were surrogates for possible rising peer challengers (a resurgent Russia, a rising China) that would justify the maintenance of large standing forces and investments in new weapon systems. Nevertheless, US forces found themselves providing troops and logistics along with international partners in various small conflicts and humanitarian crises.

The increased use and celebration of precision weapons had another effect on US military operations. The expectation grew in Western publics that the US could avoid killing civilians, even when targeting enemy forces in close proximity. And the likelihood that post-attack civilian casualties might come from infrastructure damage required US forces to limit targets during combat operations. When civilian casualties happened, many observers assumed that they were intended, no matter the official denials.

Concern about US casualties had an even greater effect on operations. There was little domestic support for many of the interventions, which meant that their cost in military lives was contentious. With an all-volunteer force and dozens of low-stake operations, force protection became the American military's highest priority. Consider US involvement in Haiti or the bombing of Serbia during the Clinton administration. The military, openly at times, was part of the political opposition to these ventures. Low or no casualties became the political trade-off both with the public and with the American military for US military interventions unclearly related to security interests.

The 1990s also marked the triumph of jointness, the idea that the services should integrate their operations. It was a particular kind of jointness that gained acceptance. The operational command structure would be joint. Some procurement would be joint. And there would be no open criticism among the services of each other's doctrines, weapon system preferences, or performance. There was collaboration for joint gain, but not to maximize organizational resources at another service's expense. To this end, there were strict, if unwritten, rules about offering performance comparisons.[6] Of course, the Air Force would occasionally boast about having the ability to win wars on its own, and the Marine Corps might tout its deployability and flexibility, but few took umbrage. These were old themes without impact on resource allocations, and all operational opportunities were being shared equitably and without visible friction. As we noted in Chapter 1, it was politicians' failure to fracture this sort of unproductive cooperation that helped the services avoid debate about which elements of military power we should protect and which we should cut. Strategy reviews such as the Base Force planning effort, the Bottom-Up Review and the Quadrennial Defense Review became kabuki theater for the protection of organizational interests. In each case the services took the reductions, and later increases, essentially evenly. Jointness in this fashion is the cartelization of the market by the producers of military power.

New challenges, unconnected answers

The services viewed the peacekeeping and humanitarian operations that produced most deployments for US forces during the 1990s as distractions from the constant training needed to keep the warrior's edge. There was no interest in becoming especially skilled in taming ethnic conflict or in keeping tabs on previously warring factions. These were seen as constabulary tasks best left to allies and, when necessary, could be easily performed by troops trained and equipped for more intense combat.

Instead, the services' visions of the future converged on "net-centric warfare" in which US superiority in RMA-type advances – numerous sensor systems covering the entire battlefield identify targets for precision attack by dispersed and networked joint forces – allowed the American military to out-know and out-fight a conventional enemy force. To be sure, the Navy version had more ships in the demonstration slideshow and the Army version included artillery and scout helicopters. But it was essentially the same PowerPoint illustration that showed the superiority of the networked and joint American military against a large enemy conventional force. It was the Cold War in very modern garb; indeed, as Robert Tomes shows in Chapter 8, this was the war such systems were developed to fight. Notably, the enemy in the PowerPoint slide was not a band of insurgents hiding among the civilian population, but rather another military fielding lots of tanks and missiles, with civilians conveniently absent from the battlefield and no vital infrastructure visibly at risk.

Neither the increase in humanitarian interventions nor the talk of RMA much affected procurement priories. All of the services worried about the renewal of their main weapon platforms, which had largely been acquired during the Reagan build-up, and which by the mid-1990s were aging in years if not in use. They invested heavily in the design and development of the next generation of fighter aircraft, naval combatants, helicopters, and armored vehicles, even though none of these weapon systems were being challenged by foreign competition, and the services' budgets were being squeezed by rising personnel and operating costs.[7] The force structure being filled was smaller and the systems more networked, but the design focus remained on obtaining weapons capable of destroying opposing militaries.

Intentionally ignored were counterinsurgency training and doctrine. The military's main lesson from Vietnam was that counterinsurgency campaigns took much longer to succeed than public support could be maintained. The so-called Powell Doctrine – Chairman Colin Powell's post-Cold War reiteration of Secretary of Defense Casper Weinberger's famous policy requirements for the use of military force – was meant to prevent involvement in precisely such efforts.[8] Its demands for the application of overwhelming force, the pursuit only of vital interests, and the announcement of exit strategies all cut against the messy and prolonged nature of counterinsurgency. There were no rewards for those who sought to nurture counterinsurgency capabilities in the military. Even the newly created Special Operations Command preferred to perfect its hostage-freeing rather than its nation-building skills.[9] No one in the military wanted to repeat the Vietnam experience, even with a better outcome.

Mostly by luck, the Clinton administration managed to avoid entangling the military in these operations. Its liberal interventionist strategy brought military capabilities to bear to stop civil wars and prevent humanitarian disasters, the kind of actions that could embroil the military in counterinsurgencies. The opposition to such interventions from Republicans and the military required the administration, especially after the Somalia experience, to limit the potential for ground combat. The Kosovo intervention, with its easy use of airpower and its great hesitancy to commit soldiers to control Kosovo and protect civilians from Serbian paramilitary and police units, illustrates this constraint. With such controls in place, the military kept its focus on the big war fight.

George W. Bush argued for a less interventionist policy during the 2000 presidential election campaign, and promised to keep the military out of nation-building efforts, exactly what the military wanted. He also promised accelerated transformation, a focus on high-technology operations. But the September 11, 2001 attacks brought US military involvement in Iraq and Afghanistan that, although they were not intended to do so, led to the low-technology, manpower-intensive nation-building and counterinsurgency operations Bush said he would avoid. There was little planning for such operations, and the forces deployed were purposely kept small to avoid them. Nevertheless, soon there was nothing else on the military's agenda. Reserves were called up and service personnel had their tours extended. The strain on recruitment, certain high-demand/low-availability specialties such as Military Police and Special Forces, ammunition stocks, and equipment was significant. And yet procurement priorities stayed focused on big-platform programs with little relevance to combat in Iraq and Afghanistan and counterterrorism.

Precision strikes and the other elements of the RMA have a role in counterterrorism and counterinsurgency campaigns, but not a big one. The image of the Special Operations sergeant in Afghanistan riding on a horse and calling in a GPS- or laser-guided weapon from a loitering B-52 or F/A-18 to destroy a Taliban or Al-Qaeda target is etched in our collective memory; so too is the account of a Predator UAV firing a missile to take out Al-Qaeda leaders in Yemen. We also know that communication satellites and infrared sensors serve US forces well in Iraq and Afghanistan. But in the end, in counterinsurgency operations, the critical parts of the effort are neighborhood informants and soldiers on patrol, with no more than some armor plate between them and a sniper's bullet or an improvised explosive. The communications that really count are between informers in the neighborhood and the security forces patrolling its streets. Technology can aid operations on the margin, but counterinsurgency requires an entirely different organizational structure and focus than the one developed to employ the RMA.

True, the services all herald their commitment to counterinsurgency operations and claim to be transforming to that end. Certainly training for such wars has increased. Fighting insurgencies naturally has affected the structure of the services, especially on the habits and skills of its future leaders. And the recent shift of funding from the Air Force and Navy to the Army and Marines along with the plan to expand the ground forces by about 10 percent shows a heightened commitment to counterinsurgency and reconstruction duties. But the shift in funding to the ground forces is small. More importantly, the procurement priorities in all the services have barely budged since the days when counterinsurgency was to be avoided as noted. The commitment to counterinsurgency is shallower than rhetoric suggests.

Explaining what changed and what didn't

The first interwar period saw significant innovations in warfare. In the 1920s and 1930s amphibious operations, the design and use of aircraft carriers, strategic bombing doctrine, and blitzkrieg were conceived or advanced. The innovations were both technological and organizational, and took place in several countries including the United States. The introduction and diffusion of the innovations came quickly in World War II.[10]

One reason that the 1920s and 1930s were militarily productive was because World War II was essentially a continuation of World War I. There were few surprises among World War II's participants and the sides they took. There were experiments with aircraft, tanks, and naval and amphibious warfare in World War I that helped show the way for crucial investments in technology in the interwar period, and the war provided important puzzles that likely had to be solved. Problems such as anti-submarine warfare, the penetration of static fronts, and the logistical support industrialized nations could offer their armies all needed to be solved. Relative to their compatriots managing militaries in other times, the allied leaders knew a great deal about the identity of the enemy and sort of forces needed to fight him. Knowledge combined with threat to create an environment conducive to innovation.

The second interwar period, the years between the Cold War and September 11, 2001, was much less supportive for would-be military innovators. Certainly, as the chapters in this volume describe, the American military did make changes after the end of the Cold War. The Army sought to become more digitalized, mobile, and networked. The US forces garrisoned in Europe and Korea shrank. Both the Air Force and the Army became more expeditionary in organizational structure and equipment. The Marine Corps embraced fads, such as nonlethal warfare and bio-terrorism attack detection. The Navy reduced the size of its fleet, cutting the submarine force in particular; improved fleet defenses; and much enhanced its ability to deliver precise airstrikes. Special Operations focused on improving internal communications and increasing its lethality.

However, the second interwar period did not produce the innovative doctrine to meet the counterinsurgency wars that absorb so much of the American military's resources and attention in the early years of the twenty-first century. Little that was done in American military training centers and defense laboratories during the years after the collapse of the Soviet Union has had much impact in the fighting that has occurred in Afghanistan and Iraq. Less obvious perhaps, but well-demonstrated by the chapters, is that the services' embrace of the RMA did not produce fundamental change in their approach to conventional war.

What explains this stasis? We have described the military as having a preference for big enemies, but we have not said why. Nor have we yet explained why the civilians who sent the military to fight a different sort of enemy failed to impose organizational changes on the military in service of these missions. What prevented military innovation since the end of the Cold War? This question is the crux of this study, but it is a bit misleading. It treats innovation as a natural result that only unfortunate conditions can stop. But innovation in large public organizations like the military services is, in fact, rare. Its presence should be the odd thing that observers struggle to explain. Its absence is what we ought to expect.

Military services are large organizations, and large organizations – especially public ones – are political arrangements. This fact makes them resist innovation for several reasons. First, within organizations the accepted way of doing business reflects a successful coalition maintained by hierarchy – a power structure linked to the status quo.[11] An organization can rearrange itself whenever it wants, in theory. But upsetting successful arrangements requires the destruction of a stable coalition, threatening the power of those that would have to upset it. Management of a large organization requires consensus, and reaffirming last year's priorities is often the fastest way to reach it.

Second, successful public organizations tend to infuse their members with a sense of purpose that makes them eager servants of the organization's ends.[12] This sense of purpose aids morale, but it means that the people in the organization are usually attached to the current way of doing business and unenthusiastic about new missions. Third, current ways of doing business have reliably brought outside support in the form of funding. Public organizations will struggle to change without the cooperation of their congressional and executive patrons.

Changing these arrangements, which innovation requires, then endangers the organization's power structure, its members' sense of purpose, and outside support. The organization naturally resists such steps. This is not to condemn leaders of these organizations or cast them as overly conservative. Doing what they have done before is their job, which is a public service. We count on federal bureaucracies for the reliable provision of public goods. Constant revolution would endanger the provision of a task that the public wants done, and remove predictability from the calculation of all those who have to deal with such organizations. Bureaucracy's virtue is in following established routines.

Moreover, innovation involves direct threats to the work and stature of specific groups, who will naturally resist. Organizational leaders will have to contend with dissent within the organization and the potential mobilization of friends outside the organization who share the groups' interests. Big changes may promise big benefits but surely also big conflicts.

To be clear, our position is not that the US military services always resist change or that it is has not made improvements since the Cold War. In general, the US military as a whole pursues, not avoids, new technology. Notwithstanding organizational lore about intransigent battleship admirals and the motor-resisting cavalry officers, the military embraces technological change. Only ill-informed commentators expect that there would be noticeable boot-dragging when it comes to employing the technologies of the RMA. As the prior chapters indicate, the services incorporated new technologies and became better at their core tasks during this period. But the reason that these organizations so resolutely pursue change that improves their ability to perform core tasks is the same reason that they do not pursue fundamental changes in those tasks. The former changes serve the organization's purpose, the latter challenge it. There should then be little surprise that the services used the technologies associated with the RMA to become more proficient at what they had always done, not to re-engineer themselves.

Of course, public organizations do occasionally innovate. They change their doctrine, their primary task, their hierarchies, and even their sense of purpose. Some of this occurred during the first interwar period. As noted in the introduction, theorists of innovation have several explanations for how public organizations innovate. Civilians, fearing threats from abroad, can empower would-be innovators within organizations to change them.[13] Shocking events, such as lost wars and scandal, can create outside demands for reform and upset dominant coalitions with organizations, producing innovation.[14] And strategies that force organizations to compete for relevance and budget share can force them to innovate in the interest of self-preservation.[15]

Two cases in this study qualify as partially innovative according to the first coding of organizational transformation – the adaptation of RMA technologies to produce organizational and doctrinal change for fighting conventional missions. These are the Army and Special Operations Forces. The presence of bureaucratic competition and some crucial civilian intervention explains this limited transformation. No case meets the criteria for the second coding of transformation – organizational and doctrinal changes for fighting the low-intensity counterinsurgency warfare the international environment produced. The absence of any innovation driving variables, and the presence of normal organizational constraints, explains the lack of innovation to fight these wars.

One can see evidence of the virtues of competition in the Kosovo campaign, wherein the Air Force's dominant role may have pushed the Army into making a greater commitment to lighter forces than otherwise would have been the case. The Army's experience in Kosovo created fears of its marginalization, driving the organization toward seeking lighter vehicles and faster deployment – in effect, attempting to replicate and exceed perceived Air Force advantages for future Kosovo-like conventional conflicts.

Similarly, perceived superior performance by the Air Force in 1991 caused technological and doctrinal change in the Navy. Naval aviation's difficulties in the Gulf War encouraged various reforms that improved carrier-based airpower. These dynamics continue today, as the Navy and Air Force argue that they are relevant to the small wars, and fight for the biggest piece of the air battle. But overall, with jointness leaving their budgets and relevance unthreatened, competitive pressures on the services are fairly weak and unlikely to produce much innovation.

The transformation of Special Operations Forces is due to civilian intervention. The Title X responsibilities granted by Goldwater-Nichols provided US SOCOM with the budgetary authority and organizational autonomy necessary to strike out on its own path.[16] Civilian strategy in Afghanistan combined with the designation of SOF as "lead" service in the War on Terror allowed the direct action units to dominate SOCOM. But because these civilian-induced changes have also been combined with lack of civilian oversight, and a certain amount of civilian ignorance of the likely tactics and consequences, it is unclear whether these changes have been salutary.

The lack of conventional transformation by the other organizations at issue, and lack of environmental transformation by all organizations surveyed, is explained by the absence of innovative variables. Civilian intervention has been weak for several reasons: partisan politics, the congressional environment, and vague and disconnected civilian leadership. Inter-service collusion has been much more dominant than competition, with predictable results. And the organizational essence and dominant coalitions within organizations has prevented intra-service change.

Partisan politics kept the national leadership weak throughout much of the second interwar period. Beset immediately by a shaky relationship with the military – brought on by issues surrounding homosexuals in the military, budget cuts, and a penchant for humanitarian intervention – the Clinton administration saw its control of military erode further with the Republican Revolution of 1994. The Republican-controlled Congress was eager to denounce the President's naiveté in meddling with the military and more inclined to defend the military's preferences than uproot them. Burdened with these difficulties and content to fight other battles, the Clinton administration set budget top lines but made little effort to alter the services' priorities.

Control of Congress gave the Bush administration greater power over the military, but its rhetorical commitment to transformation has produced little of the real thing. One problem was budgetary. By massively increasing the defense budget – it has nearly doubled, not including the direct cost of the Iraq and Afghanistan wars – soon after taking office, the administration removed the impetus to make choices among the service's procurement preferences and its own.

The Bush administration's difficulty in imposing its idea of transformation can also be explained by the fact that the idea was vague. Initially, transformation was whatever Secretary of Defense Donald Rumsfeld wanted it to mean. He seemed to use the term loosely, never offering a clear definition.[17] Later, as the Secretary's power waned, transformation was whatever the services chose it to be. Today, as the term falls into history from overuse and the Secretary's departure, transformation is everything the services and their contractors have recently fielded, have ordered, or plan to build.[18] The way the business system buys travel and keeps its books, at least on paper, is transformational, but so too are the training system, the C-17, and the new aircraft carrier. "Transformation" is a constant descriptive in military planning documents and budget justifications.

But neither the Secretary's murky words nor the services' crowded PowerPoint slides ever made reference to transformation for counterinsurgency operations or the transformation of the military's nation-building capabilities. These were the missions for which President Bush ultimately deployed the military. The Bush administration's theory of military transformation was divorced from its international strategy. It appeared to have both a vision of a high technology military future and the desire to trek through the Middle East replacing governments and restoring the irrigation systems. By creating a new set of priorities and needed reforms for the military, the wars in Afghanistan and Iraq undermined the administration's ability to find military allies to implement the conventional transformation.

Inter-service harmony also helped prevent innovation. The competition theory of innovation says that organizations can be forced into innovation if their broader sense of organizational purpose is threatened. The problem with jointness is not its operational manifestation but its ideology – the idea that contention between services is automatically bad, rather than a potential source of creative friction. Jointness, in the sense of cooperation in the budgetary process, can deprive civilians of their ability to ally with one service or community against others, undermining civilian control. If the enactment of civilian preference would bring transformation, jointness is its enemy.

Civilians, however, are not powerless in the face of service collusion. They can provoke competition among services with strategies that select among them, as with Eisenhower's New Look, which gave 50 percent of defense spending and pride of place to the Air Force, pushing the Army and Navy into reforms to recapture relevance, even at the expense of their existing doctrines.[19] The practice of giving the same share of the budget every year to each service prevents such competition and innovation.

Still another reason for the lack of innovation was a prevalent feeling of security. There were wars in the second interwar period, but as David Halberstam put it, these were wars in a time of peace. There was no imminent threat to national survival alarming civilians and military leaders alike into challenging the relevance of current doctrine. The level of interest in military affairs, even among Congressmen, was historically low during the period, leaving entities tied to the status quo as the major players attending to military reform.[20]

Even after September 11, the threat of terrorism did not induce innovation from within the military. One reason may be that counterterrorism, strictly speaking, is not a military task. The obvious reforms and innovations lie in the intelligence community, the Department of Homeland Security, and Federal Bureau of Investigation. Of course, the wars in Afghanistan and Iraq have linked counterterrorism to counterinsurgency and hence suggested possible innovation in the services. And we have seen, as noted, reforms meant to improve the military's capacity of fight these wars. But these reforms have been, and are likely to remain, shallow. The military is averse to these sorts of conflicts because they subvert its traditional purposes.

That is no accident. These wars require the sort of bloody and imperial foreign policy that Americans have historically rejected. The services are unsuited for this sort of war – in their training, procurement preferences, rotation schedules and more – because the nation has told them for decades to focus on other tasks. They hesitate to involve themselves in campaigns that may require incredible brutality to win without the public's commitment to do so. Today the nation, at least to an extent, has changed its mind, and the ground forces are shifting their training to suit this preference. But no service is shifting procurement much or talking about abandoning capability in traditional fighting.

Organizations are the products of decades of annual budget decisions and political adjustments. Routines, controls, and fixed objectives are their operational code. They do not change quickly unless the absence of change threatens the organization's survival. The current fetish for counterinsurgency expertise will aid constituencies within the services willing to specialize in such wars. But today's preference for a pacification, nation-building force will not have great impact on the services unless it is clear that the failure to change – not to be confused with mere rhetoric – will threaten the organization. So far that has not occurred. The President and Congress have not given up many surface combatants or fighters to pay for counterinsurgency capability. Meanwhile, every bloody day in Iraq and

Afghanistan tests the country's tolerance for small wars without end. It also reminds generals and admirals, many of whom never forgot, that these sorts of wars require skills in managing and policing societies that would test any military, and particularly one responsive to public opinion critical of harsh tactics. They, and we, may wonder whether these are skills worth having.

Notes

1 Tables from the 2001 Department of Defense Annual Report to Congress, demonstrating these facts, are available online at: www.dod.mil/execsec/adr2001/toc.html.
2 On American dominance see William Wohlforth, "The Stability of the Unipolar World," *International Security* (Summer, 1999); and Joseph Nye, *The Paradox of American Power: Why the World's Only Superpower Can't Go It Alone* (Oxford: Oxford University Press, 2002).
3 For an account of the infamous 1990 budget deal, and its relation to Bush's famous campaign pledge not to raise taxes, see John Robert Greene, *The Presidency of George Bush* (Lawrence, KN: University of Kansas Press, 2000), pp. 78–88.
4 From Vietnam to the end of the Cold War, the United States engaged in four modestly sized military interventions: the Dominican Republic, 1965; Lebanon, 1982; Grenada, 1983; Panama, 1989. Since the fall of the Berlin Wall, the United States has engaged in eight such interventions: the Persian Gulf War, 1991; Northern Iraq, 1991; Somalia, 1992; Haiti, 1994; Bosnia, 1995; Kosovo, 1999; Afghanistan, 2001; and Iraq, 2003. For a complete list of named US operations, see www.globalsecurity.org/military/ops/index.html.
5 See *National Defense Estimates for FY 2007 Budget* (Green Book), pp. 110–115 for RDT&E budget from 1947–2007, p. 47 for RDT&E deflators.
6 This is typical cartel behavior among public and not-for-profit organizations. See Harvey M. Sapolsky, "Organizational Competition and Monopoly," *Public Policy, the Yearbook of the Kennedy School of Government* (Cambridge, MA: Harvard University Press, 1968).
7 See Cindy Williams, ed., *Holding the Line: US Defense Alternatives for the Early Twenty-First Century* (Cambridge, MA: MIT Press, 2001); and Williams, ed., *Filling the Ranks: Transforming the US Military Personnel System* (Cambridge, MA: MIT Press, 2004).
8 See Weinberger speech, "The Uses of Military Power," delivered November 28, 1984 to the National Press Club. Reprinted in Caspar Weinberger, *Fighting for Peace: Seven Critical Years in the Pentagon* (New York, NY: Warner Books, 1990), pp. 429–445.
9 For an extended argument on this point, see Hy S. Rothstein, *Afghanistan and the Troubled Future of Unconventional Warfare* (Annapolis, MD: USNI Press, 2006).
10 For classic works on innovation, please see, *inter alia*, Barry Posen, *The Sources of Military Doctrine: France, Britain, and Germany Between the World Wars* (Ithaca, NY: Cornell University Press, 1984); Stephen Rosen, *Winning the Next War: Innovation and the Modern Military* (Ithaca, NY: Cornell University Press, 1992); and Williamson Murray and Allan Millett, *Innovation in the InterWar Period* (Cambridge, England: Cambridge University Press, 1996).
11 Rosen, *Winning the Next War.*
12 Philip Selznick, *Leadership in Administration: A Sociological Interpretation* (Berkeley, CA: University of California Press, 1957).
13 Posen, *Sources of Military Doctrine.*
14 Rosen, *Winning the Next War.*
15 Owen R. Cote, Jr., *The Politics of Innovative Military Doctrine: The US Navy and Fleet Ballistic Missiles* (PhD dissertation, Massachusetts Institute of Technology,

1996); Harvey M. Sapolsky, "On the Theory of Military Innovation," *Breakthroughs*, Vol. IX, No. 1 (Spring 2000), pp. 35–39.

16 Title X refers to the section of US Code that empowers individual military services with budgetary authority to support, train, and equip operational forces.

17 See Donald Rumsfeld, "Transforming the Military," *Foreign Affairs* (May/June 2002); and Caitlin Talmadge, "Transforming the Pentagon: McNamara, Rumsfeld, and the Politics of Change," *Breakthroughs* (2006).

18 See, for instance, the advertisements for the Army's UAV ground operating system, the Air Force's FLIR pod, and the Navy's CVN 21, in *Seapower* (April 2007).

19 Cote, *The Politics of Innovative Military Doctrine*.

20 Richard A. Lacquement, Jr., *Shaping American Military Capabilities After the Cold War* (Westport, CT: Praeger, 2003).

Index

Made in the USA
Monee, IL
14 September 2026

63198314R00120